AN ASSESSMENT OF THE SBIR PROGRAM AT THE DEPARTMENT OF ENERGY

Committee for
Capitalizing on Science, Technology, and Innovation:
An Assessment of the Small Business Innovation Research Program

Policy and Global Affairs

Charles W. Wessner, Editor

NATIONAL RESEARCH COUNCIL
OF THE NATIONAL ACADEMIES

THE NATIONAL ACADEMIES PRESS
Washington, D.C.
www.nap.edu

THE NATIONAL ACADEMIES PRESS 500 Fifth Street NW Washington, DC 20001

NOTICE: The project that is the subject of this report was approved by the Governing Board of the National Research Council, whose members are drawn from the Councils of the National Academy of Sciences, the National Academy of Engineering, and the Institute of Medicine. The members of the committee responsible for the report were chosen for their special competences and with regard for appropriate balance.

This study was supported by Contract/Grant No. DASW01-02-C-0039 between the National Academy of Sciences and U.S. Department of Defense, NASW-03003 between the National Academy of Sciences and the National Aeronautics and Space Administration, DE-AC02-02ER12259 between the National Academy of Sciences and the U.S. Department of Energy, NSFDMI-0221736 between the National Academy of Sciences and the National Science Foundation, and N01-OD-4-2139 (Task Order #99) between the National Academy of Sciences and the U.S. Department of Health and Human Services. The content of this publication does not necessarily reflect the views or policies of the Department of Health and Human Services, nor does mention of trade names, commercial products, or organizations imply endorsement by the U.S. government. Any opinions, findings, conclusions, or recommendations expressed in this publication are those of the author(s) and do not necessarily reflect the views of the organizations or agencies that provided support for the project.

International Standard Book Number-13:   978-0-309-11412-7
International Standard Book Number-10:   0-309-11412-8

Limited copies are available from the Policy and Global Affairs Division, National Research Council, 500 Fifth Street, NW, Washington, D.C. 20001; (202) 334-1529.

Additional copies of this report are available from the National Academies Press, 500 Fifth Street, NW, Lockbox 285, Washington, D.C. 20055; (800) 624-6242 or (202) 334-3313 (in the Washington metropolitan area); http://www.nap.edu.

Copyright 2008 by the National Academy of Sciences. All rights reserved.

Printed in the United States of America

THE NATIONAL ACADEMIES
Advisers to the Nation on Science, Engineering, and Medicine

The **National Academy of Sciences** is a private, nonprofit, self-perpetuating society of distinguished scholars engaged in scientific and engineering research, dedicated to the furtherance of science and technology and to their use for the general welfare. Upon the authority of the charter granted to it by the Congress in 1863, the Academy has a mandate that requires it to advise the federal government on scientific and technical matters. Dr. Ralph J. Cicerone is president of the National Academy of Sciences.

The **National Academy of Engineering** was established in 1964, under the charter of the National Academy of Sciences, as a parallel organization of outstanding engineers. It is autonomous in its administration and in the selection of its members, sharing with the National Academy of Sciences the responsibility for advising the federal government. The National Academy of Engineering also sponsors engineering programs aimed at meeting national needs, encourages education and research, and recognizes the superior achievements of engineers. Dr. Charles M. Vest is president of the National Academy of Engineering.

The **Institute of Medicine** was established in 1970 by the National Academy of Sciences to secure the services of eminent members of appropriate professions in the examination of policy matters pertaining to the health of the public. The Institute acts under the responsibility given to the National Academy of Sciences by its congressional charter to be an adviser to the federal government and, upon its own initiative, to identify issues of medical care, research, and education. Dr. Harvey V. Fineberg is president of the Institute of Medicine.

The **National Research Council** was organized by the National Academy of Sciences in 1916 to associate the broad community of science and technology with the Academy's purposes of furthering knowledge and advising the federal government. Functioning in accordance with general policies determined by the Academy, the Council has become the principal operating agency of both the National Academy of Sciences and the National Academy of Engineering in providing services to the government, the public, and the scientific and engineering communities. The Council is administered jointly by both Academies and the Institute of Medicine. Dr. Ralph J. Cicerone and Dr. Charles M. Vest are chair and vice chair, respectively, of the National Research Council.

www.national-academies.org

Committee for
Capitalizing on Science, Technology, and Innovation:
An Assessment of the Small Business Innovation Research Program

Chair
Jacques S. Gansler
Roger C. Lipitz Chair in Public Policy and Private Enterprise
and Director of the Center for Public Policy and Private Enterprise
School of Public Policy
University of Maryland

David B. Audretsch
Distinguished Professor and
 Ameritech Chair of Economic
 Development
Director, Institute for Development
 Strategies
Indiana University

Gene Banucci
Executive Chairman
ATMI, Inc.

Jon Baron
Executive Director
Coalition for Evidence-Based Policy

Michael Borrus
Founding General Partner
X/Seed Capital

Gail Cassell
Vice President, Scientific Affairs and
Distinguished Lilly Research Scholar
 for Infectious Diseases
Eli Lilly and Company

Elizabeth Downing
CEO
3D Technology Laboratories

M. Christina Gabriel
Director, Innovation Economy
The Heinz Endowments

Trevor O. Jones
Chairman and CEO
BIOMEC, Inc.

Charles E. Kolb
President
Aerodyne Research, Inc.

Henry Linsert, Jr.
Chairman and CEO
Martek Biosciences Corporation

W. Clark McFadden
Partner
Dewey & LeBoeuf, LLP

Duncan T. Moore
Kingslake Professor of Optical
 Engineering
University of Rochester

Kent Murphy
President and CEO
Luna Innovations

Linda F. Powers
Managing Director
Toucan Capital Corporation

Tyrone Taylor
President
Capitol Advisors
    on Technology, LLC

Charles Trimble
CEO, *retired*
Trimble Navigation

Patrick Windham
President
Windham Consulting

PROJECT STAFF

Charles W. Wessner
Study Director

McAlister T. Clabaugh
Program Associate

David E. Dierksheide
Program Officer

Sujai J. Shivakumar
Senior Program Officer

Jeffrey McCullough
Program Associate

RESEARCH TEAM

Zoltan Acs
University of Baltimore

Alan Anderson
Consultant

Philip A. Auerswald
George Mason University

Robert-Allen Baker
Vital Strategies, LLC

Robert Berger
Robert Berger Consulting, LLC

Grant Black
University of Indiana South Bend

Peter Cahill
BRTRC, Inc.

Dirk Czarnitzki
University of Leuven

Julie Ann Elston
Oregon State University

Irwin Feller
American Association for the Advancement of Science

David H. Finifter
The College of William and Mary

Michael Fogarty
University of Portland

Robin Gaster
North Atlantic Research

Albert N. Link
University of North Carolina

Benjamin Roberts
Harvard University

Rosalie Ruegg
TIA Consulting

Donald Siegel
University of California at Riverside

Paula E. Stephan
Georgia State University

Andrew Toole
Rutgers University

Nicholas Vonortas
George Washington University

POLICY AND GLOBAL AFFAIRS
Ad hoc Oversight Board for
Capitalizing on Science, Technology, and Innovation:
An Assessment of the Small Business Innovation Research Program

Robert M. White, Chair
University Professor Emeritus
Electrical and Computer Engineering
Carnegie Mellon University

Anita K. Jones
Lawrence R. Quarles Professor of
  Engineering and Applied Science
School of Engineering and Applied
  Science
University of Virginia

Mark B. Myers
Senior Vice President, *retired*
Xerox Corporation

Contents

PREFACE	xiii
SUMMARY	1
1 INTRODUCTION	11

    1.1   SBIR Creation and Assessment, 11
    1.2   SBIR Program Structure, 12
    1.3   SBIR Reauthorizations, 13
    1.4   Structure of the NRC Study, 14
    1.5   SBIR Assessment Challenges, 15
    1.6   Assessing SBIR at the Department of Energy (DoE), 19
           1.6.1   Surveys of DoE SBIR Award-recipient Companies, 19
           1.6.2   Case Studies, 21
    1.7   Structure of the Report, 24

2 FINDINGS AND RECOMMENDATIONS	25
3 AWARD STATISTICS	42

    3.1   Trends in Energy Research and Development, 43
    3.2   Size of Individual Awards, 45
           3.2.1   Phase I Awards, 46
           3.2.2   Phase II Awards, 47
    3.3   Geographic Concentration, 47

3.4 Multiple-Award Winners, 50
    3.4.1 SBIR Award Clustering to Support Technology Development, 52
    3.4.2 Development Funding Prior to SBIR Award, 54

4 COMMERCIALIZATION   **56**
4.1 Challenges of Commercialization, 56
4.2 Project Status, 57
    4.2.1 Project Discontinuation, 58
4.3 Sales and Licensing, 58
    4.3.1 Skew Effects, 59
    4.3.2 Sales Expectations and Likely Future Sales, 59
    4.3.3 Licensing, 61
    4.3.4 Customers, 61
    4.3.5 Marketing, 62
    4.3.6 Additional Development Funding, 62
4.4 Further Investment: Phase III at DoE, 64
    4.4.1 DoE SBIR and Venture Capital (VCs), 64
    4.4.2 Equity Investments from Large Corporations, 65
    4.4.3 Other Resources, 65
    4.4.4 Matching Funds and Cost-sharing, 66
    4.4.5 Non-SBIR Federal Funding, 66
4.5 Employment Effects, 66
4.6 Phase I Commercialization, 67
    4.6.1 Commercialization Resulting from the Phase I Projects, 67
    4.6.2 Follow-on Development Funding Resulting from the Phase I Projects, 68
    4.6.3 Other Benefits of Phase I-only Projects, 69
4.7 Multiple-Award Winners, 70

5 AGENCY MISSION   **72**
5.1 Managing a Program with Multiple Objectives, 72
5.2 Alignment Issues for SBIR and the DoE Mission, 74
    5.2.1 Research vs. Commercial Culture, 74
    5.2.2 SBIR as a Tax, 74
    5.2.3 Administrative Burdens, 74
5.3 Changing Perceptions of SBIR, 75
    5.3.1 Supporting Program Missions, 75
    5.3.2 Providing Research Quality, 75
    5.3.3 Research Impact, 76
    5.3.4 Comparative Research Value, 77
    5.3.5 Project Ownership, 77

5.4 Capitalizing on Program Flexibility, 78
    5.4.1 Balancing Commercialization and Mission Orientation, 78
    5.4.2 Internal Reallocation of Topics Among Programs, 79

6 WOMAN- AND MINORITY-OWNED BUSINESSES   **80**
6.1 Woman-owned Businesses, 80
6.2 Minority-owned Businesses, 81
6.3 Success Rates for the Different Groups, 82

7 KNOWLEDGE EFFECTS   **85**
7.1 Publications and Intellectual Property, 85
7.2 Stimulating New Research, 86
7.3 Building Partnerships and Enhancing Networks, 87
7.4 SBIR and the Universities, 88

8 PROGRAM MANAGEMENT   **90**
8.1 SBIR in the Department of Energy, 90
8.2 Resources for Program Administration, 93
8.3 Topic Generation, 94
8.4 Award Selection, 95
    8.4.1 First-step Technical Review, 95
    8.4.2 Initial Review Approaches, 96
    8.4.3 1995 Process Revisions, 97
    8.4.4 Fairness of Competition, 97
8.5 Outreach, 98
8.6 The Application and Award Process: Awardee Comments, 98
8.7 Managing Information on Awards, 99
    8.7.1 Reporting Requirements, 99
    8.7.2 Freedom of Information Act, 99
8.8 Program Structure, 100
    8.8.1 Differences Between Agencies, 100
    8.8.2 Award Limits, 100
    8.8.3 Time Frames, 100
    8.8.4 Gaps between SBIR Phase I and Phase II Funding, 101
8.9 Participation of DoE National Laboratories in SBIR, 101
    8.9.1 Overview of DoE National Laboratories, 101
    8.9.2 Why SBIR Collaborations Are Not More Frequent, 102
8.10 Developments in Program Administration Since 2003, 103
    8.10.1 Online Capabilities and Plans, 103
    8.10.2 Program Manager Given More Control, 104
    8.10.3 Phase II Supplemental Awards, 104

8.11 Actions Taken by DoE SBIR Program to Encourage Commercialization, 105
    8.11.1 Evidence of Commercialization Included in Phase II Criteria, 105
    8.11.2 Commercialization Assistance Services for SBIR Awardees, 105
    8.11.3 Collecting Phase III Data, 108
    8.11.4 Recognizing Success, 110

APPENDIXES

A	**DOE SBIR PROGRAM DATA**	**113**
B	**NRC PHASE II SURVEY**	**135**
C	**NRC PHASE I SURVEY**	**155**
D	**CASE STUDIES**	**165**

    Airak, Inc., 165
    Atlantia Offshore Limited, 170
    Creare, Inc., 176
    Diversified Technologies, Inc., 185
    Eltron Research, Inc., 193
    IPIX, Inc., 199
    NanoSonic, Inc., 204
    NexTech Materials, Inc., 209
    Princeton Polymer Laboratories, Inc., 216
    Thunderhead Engineering, 221

E	**BIBLIOGRAPHY**	**227**

Preface

Today's knowledge economy is driven in large part by the nation's capacity to innovate. One of the defining features of the U.S. economy is a high level of entrepreneurial activity. Entrepreneurs in the United States see opportunities and are willing and able to take on risk to bring new welfare-enhancing, wealth-generating technologies to the market. Yet, while innovation in areas such as genomics, bioinformatics, and nanotechnology present new opportunities, converting these ideas into innovations for the market involves substantial challenges.[1] The American capacity for innovation can be strengthened by addressing the challenges faced by entrepreneurs. Public-private partnerships are one means to help entrepreneurs bring new ideas to market.[2]

The Small Business Innovation Research (SBIR) program is one of the largest examples of U.S. public-private partnerships. Founded in 1982, SBIR was designed to encourage small business to develop new processes and products and to provide quality research in support of the many missions of the U.S. government. By including qualified small businesses in the nation's R&D effort, SBIR grants are intended to stimulate innovative new technologies to help agencies meet the specific research and development needs of the nation in many areas, including health, the environment, and national defense.

---

[1] See Lewis M. Branscomb, Kenneth P. Morse, Michael J. Roberts, and Darin Boville, *Managing Technical Risk: Understanding Private Sector Decision Making on Early-Stage Technology Based Projects,* Washington, D.C.: National Institute of Standards and Technology, 2000.

[2] For a summary analysis of best practice among U.S. public-private partnerships, see National Research Council, *Government-Industry Partnerships for the Development of New Technologies: Summary Report,* Charles W. Wessner, ed., Washington, D.C.: The National Academies Press, 2002.

As the SBIR program approached its twentieth year of operation, the U.S. Congress asked the National Research Council to conduct a "comprehensive study of how the SBIR program has stimulated technological innovation and used small businesses to meet federal research and development needs" and make recommendations on still further improvements to the program.[3] To guide this study, the National Research Council drew together an expert committee that included eminent economists, small businessmen and women, and venture capitalists. The membership of this committee is listed in the front matter of this volume. Given the extent of 'green-field research' required for this study, the Steering Committee in turn drew on a distinguished team of researchers to, among other tasks, administer surveys and case studies, and to develop statistical information about the program. The membership of this research team is also listed in the front matter to this volume.

This report is one of a series published by the National Academies in response to the congressional request. The series includes reports on the Small Business Innovation Research Program at the Department of Defense (DoD), the Department of Energy (DoE), the National Aeronautics and Space Administration (NASA), the National Institutes of Health (NIH), and the National Science Foundation (NSF)—the five agencies responsible for 96 percent of the program's operations. It includes, as well, an Overview Report that provides assessment of the program's operations across the federal government, based on the assessments of the SBIR program at each of the five agencies. Other reports in the series include a summary of the 2002 conference that launched the study, and a summary of the 2005 conference on *SBIR and the Phase III Challenge of Commercialization* that focused on the Department of Defense and NASA.

PROJECT ANTECEDENTS

The current assessment of the SBIR program follows directly from an earlier analysis of public-private partnerships by the National Research Council's Board on Science, Technology, and Economic Policy (STEP). Under the direction of Gordon Moore, Chairman Emeritus of Intel, the NRC Committee on Government Industry Partnerships prepared eleven volumes reviewing the drivers of cooperation among industry, universities, and government; operational assessments of current programs; emerging needs at the intersection of biotechnology and information technology; the current experience of foreign government partnerships and opportunities for international cooperation; and the changing roles of government laboratories, universities, and other research organizations in the national innovation system.[4]

---

[3]See SBIR Reauthorization Act of 2000 (H.R. 5667—Section 108).

[4]For a summary of the topics covered and main lessons learned from this extensive study, see National Research Council, *Government-Industry Partnerships for the Development of New Technologies: Summary Report,* op. cit.

This analysis of public-private partnerships included two published studies of the SBIR program. Drawing from expert knowledge at a 1998 workshop held at the National Academy of Sciences, the first report, *The Small Business Innovation Research Program: Challenges and Opportunities,* examined the origins of the program and identified some operational challenges critical to the program's future effectiveness.[5] The report also highlighted the relative paucity of research on this program.

Following this initial report, the Department of Defense asked the NRC to assess the Department's Fast Track Initiative in comparison with the operation of its regular SBIR program. The resulting report, *The Small Business Innovation Research Program: An Assessment of the Department of Defense Fast Track Initiative,* was the first comprehensive, external assessment of the Department of Defense's program. The study, which involved substantial case study and survey research, found that the SBIR program was achieving its legislated goals. It also found that DoD's Fast Track Initiative was achieving its objective of greater commercialization and recommended that the program be continued and expanded where appropriate.[6] The report also recommended that the SBIR program overall would benefit from further research and analysis, a perspective adopted by the U.S. Congress.

SBIR REAUTHORIZATION AND CONGRESSIONAL REQUEST FOR REVIEW

As a part of the 2000 reauthorization of the SBIR program, Congress called for a review of the SBIR programs of the agencies that account collectively for 96 percent of program funding. As noted, the five agencies meeting this criterion, by size of program, are the Department of Defense, The National Institutes of Health, the National Aeronautics and Space Administration, the Department of Energy, and the National Science Foundation.

Congress directed the NRC, via HR 5667, to evaluate the quality of SBIR research and evaluate the SBIR program's value to the agency mission. It called for an assessment of the extent to which SBIR projects achieve some measure of commercialization, as well as an evaluation of the program's overall economic and noneconomic benefits. It also called for additional analysis as required to support specific recommendations on areas such as measuring outcomes for agency strategy and performance, increasing federal procurement

---

[5] See National Research Council, *The Small Business Innovation Research Program: Challenges and Opportunities,* Charles W. Wessner, ed., Washington, D.C.: National Academy Press, 1999.

[6] See National Research Council, *The Small Business Innovation Research Program: An Assessment of the Department of Defense Fast Track Initiative,* Charles W. Wessner, ed., Washington, D.C.: National Academy Press, 2000. Given that virtually no published analytical literature existed on SBIR, this Fast Track study pioneered research in this area, developing extensive case studies and newly developed surveys.

of technologies produced by small business, and overall improvements to the SBIR program.[7]

ACKNOWLEDGMENTS

On behalf of the National Academies, we express our appreciation and recognition for the insights, experiences, and perspectives made available by the participants of the conferences and meetings, as well as survey respondents and case study interviewees who participated over the course of this study. We are also very much in debt to officials from the leading departments and agencies. Among the many who provided assistance to this complex study, for this volume, we are especially in debt to Larry James of the Department of Energy and Robert Berger, formerly of the Department of Energy.

The Committee's research team deserves recognition for their instrumental role in the preparation and many revisions of this report. In that regard, special thanks are due to Philip Auerswald of George Mason University, Nicholas Vonortas of George Washington University, Grant Black of Georgia State University, and for the report's completion, Robin Gaster of North Atlantic Research Inc. Without their collective efforts, amidst many other competing priorities, it would not have been possible to prepare this report. Among the many contributing Committee members, special thanks are due to Charles Kolb of Aerodyne Research.

NATIONAL RESEARCH COUNCIL REVIEW

This report has been reviewed in draft form by individuals chosen for their diverse perspectives and technical expertise, in accordance with procedures approved by the National Academies' Report Review Committee. The purpose of this independent review is to provide candid and critical comments that will assist the institution in making its published report as sound as possible and to ensure that the report meets institutional standards for objectivity, evidence, and responsiveness to the study charge. The review comments and draft manuscript remain confidential to protect the integrity of the process.

We wish to thank the following individuals for their review of this report: David Bodde, Clemson University; George Eads, CRA International; Maxine Savitz (Retired), Honeywell, Inc.; and Roland Tibbets, SEARCH Corporation.

Although the reviewers listed above have provided many constructive comments and suggestions, they were not asked to endorse the conclusions or

---

[7]Chapter 3 of the Committee's Methodology Report describes how this legislative guidance was drawn out in operational terms. National Research Council, *An Assessment of the Small Business Innovation Research Program—Project Methodology*, Washington, D.C.: The National Academies Press, 2004. Access this report at <http://www7.nationalacademies.org/sbir/SBIR_Methodology_Report.pdf>.

recommendations, nor did they see the final draft of the report before its release. The review of this report was overseen by Robert Frosch, Harvard University, and Robert White, Carnegie Mellon University. Appointed by the National Academies, they were responsible for making certain that an independent examination of this report was carried out in accordance with institutional procedures and that all review comments were carefully considered. Responsibility for the final content of this report rests entirely with the authoring committee and the institution.

Jacques S. Gansler   Charles W. Wessner

Summary

I. INTRODUCTION

The Small Business Innovation Research (SBIR) program was created in 1982 through the Small Business Innovation Development Act. As the SBIR program approached its twentieth year of operation, the U.S. Congress requested the National Research Council (NRC) of the National Academies to "conduct a comprehensive study of how the SBIR program has stimulated technological innovation and used small businesses to meet Federal research and development needs" and to make recommendations with respect to the SBIR program. Mandated as a part of SBIR's reauthorization in late 2000, the NRC study has assessed the SBIR program as administered at the five federal agencies that together make up some 96 percent of SBIR program expenditures. The agencies, in order of program size are the Department of Defense (DoD), the National Institutes of Health (NIH), the National Aeronautics and Space Administration (NASA), the Department of Energy (DoE), and the National Science Foundation (NSF).

Based on that legislation, and after extensive consultations with both Congress and agency officials, the NRC focused its study on two overarching questions.[1]

---

[1] Three primary documents condition and define the objectives for this study: These are the Legislation—H.R. 5667, the NAS-Agencies *Memorandum of Understanding,* and the NAS contracts accepted by the five agencies. These are reflected the Statement of Task addressed to the Committee by the Academies leadership. Based on these three documents, the NRC Committee developed a comprehensive and agreed set of practical objectives to be reviewed. These are outlined in the Committee's formal Methodology Report, particularly Chapter 3, "Clarifying Study Objectives." National Research Council, An Assessment of the Small Business Innovation Research Program: Project Methodology, Washington, D.C.: The National Academies Press, 2004. Accessed at <http://books.nap.edu/catalog.php?record_id=11097#toc>.

First, how well do the agency SBIR programs meet four societal objectives of interest to Congress: (1) to stimulate technological innovation; (2) to increase private sector commercialization of innovations; (3) to use small business to meet federal research and development needs; and (4) to foster and encourage participation by minority and disadvantaged persons in technological innovation.[2] Second, can the management of agency SBIR programs be made more effective? Are there best practices in agency SBIR programs that may be extended to other agencies' SBIR programs?

To satisfy the congressional request for an external assessment of the program, the NRC conducted empirical analyses of the operations of SBIR based on commissioned surveys and case studies. Agency-compiled program data, program documents, and the existing literature were reviewed. In addition, extensive interviews and discussions were conducted with program managers, program participants, agency 'users' of the program, as well as program stakeholders.

The study as a whole sought to answer questions of program operation and effectiveness, including the quality of the research projects being conducted under the SBIR program, the commercialization of the research, and the program's contribution to accomplishing agency missions. (See Box S-1, which highlights findings from the case studies on how SBIR companies commercialize.) To the extent possible, the evaluation included estimates of the benefits (both economic and noneconomic) achieved by the SBIR program, as well as broader policy issues associated with public-private collaborations for technology development and government support for high technology innovation.

Taken together, this study is the most comprehensive assessment of SBIR to date. Its empirical, multifaceted approach to evaluation sheds new light on the operation of the SBIR program in the challenging area of early stage finance. As with any assessment, particularly one across five quite different agencies and departments, there are methodological challenges. These are identified and discussed at several points in the text.[3] This important caveat notwithstanding, the scope and diversity of the report's research should contribute significantly to the understanding of the SBIR program's multiple objectives, measurement issues, operational challenges, and achievements. This volume presents the Committee's assessment of the SBIR program at the Department of Energy.

---

[2]These congressional objectives are found in the Small Business Innovation Development Act (PL 97-219). In reauthorizing the program in 1992 (PL 102-564), Congress expanded the purposes to "emphasize the program's goal of increasing private sector commercialization developed through federal research and development and to improve the federal government's dissemination of information concerning small business innovation, particularly with regard to woman-owned business concerns and by socially and economically disadvantaged small business concerns."

[3]See, for example, Box 1-1 in Chapter 1 of this report, which lists the multiple sources of bias found in large innovation surveys.

> **Box S-1**
> **How SBIR Companies Commercialize—**
> **Findings from the Case Studies**
>
> Companies interviewed for this study illustrate some of the many approaches to commercialization taken by DoE SBIR recipient firms. The case studies of these companies can be found in Appendix D.
> One company marketed the technology developed with its lone project, sub-contracted production, and achieved huge revenues (Atlantia). Another marketed and manufactured its own product (IPIX), eventually going public. Three companies are manufacturing products based on their SBIR work, while they seek larger partners in order to expand markets (NanoScience, NexTech, Thunderhead Engineering). Other companies achieve commercialization by spinning off (Creare) or licensing (Eltron) their SBIR technologies. Yet another used SBIR research to create a new market—in one instance for its core technology (Diversified Technologies Inc.) and, in another, for an R&D contracting business (PPL).
> Although some case study companies are currently dependent on SBIR as a major source of their revenues (Airak, Creare, NanoScience), all are actively engaged in commercialization, selling either products or services. Most companies use patents to protect their intellectual property.

II. SBIR AT THE DEPARTMENT OF ENERGY

The DoE's SBIR program is located administratively within its Office of Science (SC). The Office of Science funds research in the basic energy sciences, biological and environmental research, fusion energy sciences, high energy and nuclear physics, and computational science. With an annual budget of approximately $3.5 billion, the Office of Science is the largest federal sponsor of materials and chemical sciences research within the federal government. And with a budget of $104 million in FY2005, the DoE SBIR program is the nation's third largest. (See Figure S-1.)

Given that federal agencies are barred from using any of their SBIR budgets to fund the administrative costs of the program, DoE officials often regard SBIR administrative expenses as an additional, though unspecified, tax.[4] As host for the SBIR office, DoE's Office of Science is responsible for the direct costs of administering the program: salaries for the federal employees, support services contracts, and costs such as developing and maintaining the electronic grant management system.

Because the Office of Science must use its own funds to administer the SBIR program, and because of the historical resistance to SBIR at DoE, there has been a tendency for the Office of Science to limit the resources for administering the

---
[4]Public Law 102-564.

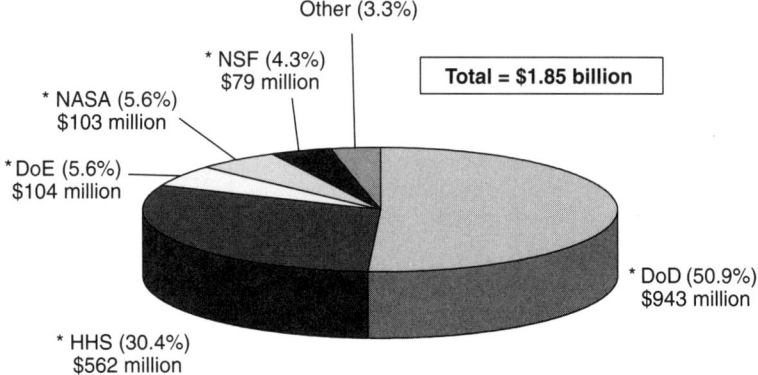

FIGURE S-1 Dimensions of the SBIR program in 2005.
SOURCE: U.S. Small Business Administration. Accessed at <*http://tech-net.sba.gov/*>.

program—ultimately to the point where 1.5 full-time DoE employees, assisted by five contractors, were managing the administration (though not the review) of approximately 1,500 applications annually (both Phase I and Phase II). In March 2005, DoE SBIR reported a staff of three federal employees and six contractors. They still reported substantial resource constraints, including the use of 20-25 percent of their time on redundant paperwork issues.

In order to cope with these resource constraints, the SBIR staff has placed a high premium on sticking to a clearly mapped-out schedule and meeting established deadlines. Low funding levels for administration mean however that the DoE SBIR staff devotes nearly all their time to managing the processes for generating technical topics, and for receiving, evaluating, and selecting grant applications. This leaves little time for activities such as outreach, measuring Phase III activity, encouraging Phase III activity (both within and outside the Department, including the national laboratories), internal evaluation, strategic planning, and the documentation of successes.

Despite these resource constraints, the NRC committee found that the Department of Energy has made significant progress in meeting the legislative objectives of SBIR and that the program is addressing the mission of the Department of Energy. As noted below, the committee has also identified a set of related recommendations that are designed to improve the operation of the SBIR program at the Department of Energy.

III. SUMMARY OF KEY PROGRAM FINDINGS

The DoE SBIR program is making significant progress in achieving the congressional goals for the program. The SBIR program is sound in concept and

> **Box S-2**
> **The Mission of the Department of Energy**
>
> DoE's mission is "to advance the national, economic, and energy security of the United States; to promote scientific and technological innovation in support of that mission; and to ensure the environmental cleanup of the national nuclear weapons complex." The agency has identified four strategic goals that support achieving this mission:
>
> **Defense Strategic Goal.** Protecting national security by applying advanced science and nuclear technology to the nation's defense.
>
> **Energy Strategic Goal.** Protecting national and economic security by promoting a diverse supply and delivery of reliable, affordable, and environmentally sound energy.
>
> **Science Strategic Goal.** Protecting national and economic security by providing world-class scientific research capacity and advancing scientific knowledge.
>
> **Environment Strategic Goal.** Protecting the environment by providing a responsible resolution to the environmental legacy of the Cold War and by providing for the permanent disposal of the nation's high-level radioactive waste.
>
> SOURCE: Department of Energy.

effective in practice at DoE. With the programmatic changes recommended by the Committee in this report, the SBIR program should be even more effective in achieving its legislative goals. Overall, the program has made significant progress in achieving its congressional objectives by:

- **Stimulating technological innovation.**[5]
  - **Generating patents and publications**. The SBIR program at DoE supports knowledge transfer in several ways. A significant number of the projects responding to the NRC Phase II Survey (43 percent) reported at least one patent application. Of these, over a third (37 percent) reported receiving a patent related to the SBIR project. In addition, nearly half (or 46 percent) of projects surveyed by NRC resulted in at least one peer-reviewed article.[6]

---

[5] See Finding G on "Knowledge Creation and Dissemination Effects" in Chapter 2.

[6] Without detailed identifying data on these patents and publications, it is not feasible to apply bibliometric and patent analysis techniques to assess the relative importance of these patents and publications.

- o **Stimulating the transfer of technology from universities to the market.** The NRC Phase II survey also suggests that SBIR awards are supporting the transfer of knowledge, firm creation, and partnerships between universities and the private sector:[7] At DoE, just over one-third of projects had some alignment with a university, through the use of university faculty as contractors on the project, use of universities as subcontractors, or employment of graduate students
- o **Indirect paths.** Case studies conducted for this study provide anecdotal evidence concerning beneficial "indirect path" effects—that projects provide investigators and research staff with knowledge that may later become relevant in a different context—often in another project or even another company.

- **Using small businesses to meet federal research and development needs.**[8]
  - o **Mission alignment.** The DoE SBIR program is operated in alignment with the department's mission. SBIR awards at DoE appear to be selected primarily on the basis of their potential to advance knowledge and provide solutions in the field of energy. There is also no evidence that DoE awards are made in fields not directly linked to the department's mission.[9]
  - o **Potential for collaboration with the National Laboratories has not been fully realized.** The National Laboratories, while an important source of technical reviewers for SBIR proposals, are not strongly involved in the SBIR program. As a result, the potentially significant role of the National Laboratories as partners with SBIR-award-recipient firms does not appear to be fully realized.
  - o **Encouraging small business formation.** The SBIR program at DoE has provided significant support for small business, frequently acting as the impetus for the foundation of new firms. Nearly one-fourth of survey respondents with DoE awards indicated that their companies were founded entirely or partly because of an SBIR award.
  - o **Providing a catalyst for research.** SBIR awards encourage research projects that would not otherwise have been undertaken. More than 80 percent of DoE NRC Phase II Survey respondents reported that they

---

[7] See Table 4-14 in National Research Council, *An Assessment of the Small Business Innovation Research Program*, Charles W. Wessner, ed., Washington, D.C.: The National Academies Press, 2008.

[8] See related Findings C and D in Chapter 2.

[9] Among case study firms, networking was identified as critical when attempting to find partners to assist with commercialization. Possibly the most dramatic identified success from the program has been Atlantia's SeaStar® technology, a cost-effective alternative for the development of otherwise inaccessible small oil fields in deep water. The technology generated more than $500 million in revenues for Atlantia.

would likely or certainly not have gone forward with their project in the absence of SBIR funding.[10]
- o **Partnering and networking**. SBIR further facilitates technological innovation through the creation of new partnerships and the strengthening of networks of innovators.[11] Innovative small businesses also use SBIR to fund alternative development strategies, exploring technological options in parallel with other activities.
- o **Certification effect**. Receipt of SBIR awards can, in addition, provide a "stamp of approval," allowing smaller companies to better access large private sector companies, including government contractors, and attract private investors.[12]

- **Fostering and encouraging participation by minorities and disadvantaged persons in technological innovation.**[13]
  - o **Consistent support, but no long-term trend is apparent**. Data from SBA indicates that between 1992 and 2005, the Phase I share of woman- and minority-owned businesses at DoE averaged just over 20 percent, with a decline in the early years of this decade, followed by an increase, accelerating in 2005. (See Figure 2-1.) Data for Phase II are similar, although slightly lower. On average, woman- and minority-owned firms won 22.1 percent of Phase I awards from 2001-2005, and 19.1 percent of Phase II awards.[14]
  - o **Lagging application success rates**. For Phase I, applications from woman-owned businesses have had a lower rate of success compared to all other groups—by approximately 3 to 10 percentage points—in every year except one. For minority-owned companies, the success rate is better than for woman-owned companies, but still lags behind the "other" category (neither woman-owned nor minority-owned). For 2002-2003, the success rate of minority-owned businesses was considerably lower than that for woman-owned and all other businesses.[15]

- **Increasing private-sector commercialization derived from federal research and development.**[16]

---

[10]See Figure 7-1.
[11]See Figure 7-2.
[12]For an early reference to the certification effect, see Joshua Lerner, "Public Venture Capital: Rationales and Evaluation," in National Research Council, *The Small Business Innovation Research Program: Challenges and Opportunities*, Charles W. Wessner, ed., Washington, D.C.: National Academy Press, 1999.
[13]See related Finding E in Chapter 2.
[14]See Chapter 6.
[15]See Chapter 6.
[16]See related Finding B in Chapter 2.

o **Significant commercialization.** A significant percentage of DoE SBIR projects are commercialized to some degree. The NRC Phase II survey data indicate that 41 percent of SBIR-funded projects reach the marketplace or have commercialization underway.[17] Of the DoE SBIR award recipient firms that responded to the NRC Phase II Survey and reported sales of some type, 76 percent sold to domestic private sector firms and 14 percent to export markets.[18]

The NRC Phase II Survey data also show that a much smaller number (4 percent) of projects generate more than $5 million in revenues.[19] However, as in cases where the market is inherently limited—as with sensitive energy technologies—products developed with DoE SBIR assistance often cannot become large commercial successes.[20]

o **Commercialization support.** DoE has a strong and innovative history of commercialization support, and has expanded its efforts in recent years. DoE has sponsored a Commercialization Assistance Program (CAP) for the past 17 years. That program has provided, on a voluntary basis to Phase II awardees, individual assistance in developing business plans and in preparation of presentations to potential investors. For the past 6 years, DoE has also offered its Phase II awardees additional market identification and networking services.[21] These programs provide professional assistance in business plan development and/or market evaluation to participating SBIR companies.

IV. SUMMARY OF KEY PROGRAM RECOMMENDATIONS

The NRC committee's recommendations are designed to improve the already effective operation of the SBIR program at the Department of Energy. They complement the core findings that the program is addressing its legislative goals. With the programmatic changes recommended here, the SBIR program should be even more effective in achieving its legislative goals.

- **Improve program processes.**
    o **Provide pre-release information on topics.** Pre-proposal communication through a fair, transparent mechanism is likely to improve proposal quality and, therefore, overall program effectiveness. The

---

[17]See Figure 4-1. Twenty-four percent of NRC Phase II Survey respondents reported products/services/or processes in use at the time of the survey, and 18 percent reported commercialization underway (figures rounded). See NRC Phase II Survey, question 1.

[18]See Table 4-2.

[19]See NRC Phase II Survey, Question 4b.

[20]The case study of Diversified Technologies illustrates this phenomenon with regard to specialized transformers. See Appendix D.

[21]DoE SBIR publications and Web site; interviews with DoE SBIR staff.

Department of Energy might consider the DoD Pre-Release information exchange model, whereby the relevant technical officer for each topic is available via email and phone for questions during a period before the official release of the solicitation.[22]

o **Develop a match-making function for SBIR awardees.**[23] The department's SBIR program should bring SBIR participants together with potential corporate customers, perhaps in trade show, technical challenge workshop, or technology demonstration/validation formats. These functions could include large corporations identified by the agency's two SBIR commercialization assistance vendors, including large energy technology corporations that serve as DoE contractors.

o **Rationalize funding allocations.**[24] The department might consider whether current ad hoc funding allocations—based on the origins of the SBIR funding by program rather than according to project quality and need—are sufficient, or whether a policy that allows individual DoE programs to benefit from SBIR projects in excess of their contributed amounts would prove more effective.

- **Encourage collaboration with National Laboratories.**[25]
  o DoE should encourage the National Laboratories to participate as subcontractors to the small business on SBIR projects, not least by removing regulatory barriers to the use of National Laboratories as subcontractors. DoE should also develop procedures to track the relationship between National Laboratories and the SBIR program more formally, including the documentation of Phase III successes.

- **Increase participation by woman- and minority-owned firms.**[26]
  o DoE should undertake an assessment of the participation rates of woman- and minority-owned firms in its SBIR program, and identify strategies to improve their success rates. The development of outreach efforts and other strategies should be based on empirical analysis of past proposals and feedback from the affected groups.[27]

---

[22]See related Recommendation A in Chapter 2. Since 2006, the relevant DoE technical officer has been available via email up to the closing date of the announcement.
[23]See related Recommendation B in Chapter 2.
[24]See related Recommendation C in Chapter 2.
[25]See related Recommendation D in Chapter 2.
[26]See related Recommendation E in Chapter 2.
[27]This recommendation should not be interpreted as lowering the bar for the acceptance of proposals from woman- and minority-owned companies, but rather as assisting them to become able to meet published criteria for grants at rates similar to other companies on the basis of merit, and to ensure that there are no negative evaluation factors in the review process that are biased against these groups.

- **Produce enhanced reports on the program for DoE management and the Congress.**
    - o DoE should provide Congress with a summary annual report on the SBIR program. This should include descriptive statistics for applications, awards, and outcomes along the dimensions identified in this report, including knowledge creation, technology innovation, and impact on agency mission, as well as commercialization.

- **Conduct regular internal and external assessments.**[28]
    - o As part of this assessment process, DoE should produce regular reports from the commercialization database. In addition to improved internal assessment capabilities that can be used to enhance program operations, DoE should also commission regular external arms-length evaluations to assess the program progress.

- **Consider the creation of an advisory board.**[29]
    - o DoE should consider the creation of an independent advisory board that draws together senior agency management, SBIR managers, and other stakeholders as well as outside experts to review current operations and achievements and, as appropriate, recommend changes to the SBIR program.

- **Provide additional management funding.**[30]
    - o Managing what should be a data-driven program requires high quality data and systematic assessment, which in turn requires appropriate funding. Increased funding is also needed to provide effective oversight, including site visits, program review, systematic third-party assessments, and other necessary management activities.

---

[28] See related Recommendation F in Chapter 2.
[29] See related Recommendation H in Chapter 2.
[30] See related Recommendation G in Chapter 2.

1

Introduction

1.1 SBIR CREATION AND ASSESSMENT

Created in 1982 by the Small Business Innovation Development Act, the Small Business Innovation Research (SBIR) program was designed to stimulate technological innovation among small private-sector businesses while providing the government cost-effective new technical and scientific solutions to challenging mission problems. SBIR was also designed to help to stimulate the U.S. economy by encouraging small businesses to market innovative technologies in the private sector.[1]

As the SBIR program approached its twentieth year of existence, the U.S. Congress requested that the National Research Council (NRC) of the National Academies conduct a "comprehensive study of how the SBIR program has stimulated technological innovation and used small businesses to meet Federal research and development needs," and make recommendations on improvements to the program.[2] Mandated as a part of SBIR's renewal in late 2000, the NRC study has assessed the SBIR program as administered at the five federal agencies that together make up 96 percent of SBIR program expenditures. The agencies are, in decreasing order of program size: the Department of Defense (DoD), the

---

[1] The SBIR legislation drew from a growing body of evidence, starting in the late 1970s and accelerating in the 1980s, which indicated that small businesses were assuming an increasingly important role in both innovation and job creation. This evidence gained new credibility with empirical analysis by Zoltan Acs and David Audretsch of the U.S. Small Business Innovation Database, which confirmed the increased importance of small firms in generating technological innovations and their growing contribution to the U.S. economy. See Zoltan Acs and David Audretsch, *Innovation and Small Firms*, Cambridge, MA: The MIT Press, 1990.

[2] See Public Law 106-554, Appendix I—H.R. 5667, Section 108.

National Institutes of Health (NIH), the National Aeronautics and Space Administration (NASA), the Department of Energy (DoE), and the National Science Foundation (NSF).

The NRC Committee assessing the SBIR program was not asked to consider if SBIR should exist or not—Congress has affirmatively decided this question on three occasions.[3] Rather, the Committee was charged with providing assessment-based findings to improve public understanding of the operations, achievements, and challenges of the program as well as recommendations to improve the program's effectiveness.

1.2 SBIR PROGRAM STRUCTURE

Eleven federal agencies are currently required to set aside 2.5 percent of their extramural research and development budget exclusively for SBIR contracts. Each year these agencies identify various R&D topics, representing scientific and technical problems requiring innovative solutions, for pursuit by small businesses under the SBIR program. These topics are bundled together into individual agency "solicitations"—publicly announced requests for SBIR proposals from interested small businesses. A small business can identify an appropriate topic it wants to pursue from these solicitations and, in response, propose a project for an SBIR grant. The required format for submitting a proposal is different for each agency. Proposal selection also varies, though peer review of proposals on a competitive basis by experts in the field is typical. Each agency then selects the proposals that are found best to meet program selection criteria, and awards contracts or grants to the proposing small businesses.

As conceived in the 1982 Act, SBIR's grant-making process is structured in three phases:

- Phase I grants essentially fund feasibility studies in which award winners undertake a limited amount of research aimed at establishing an idea's scientific and commercial promise. Today, the legislative guidance anticipates normal Phase I grants around $100,000.[4]
- Phase II grants are larger—typically about $750,000—and fund more extensive R&D to develop the scientific and commercial promise of research ideas.
- Phase III. During this phase, companies do not receive further SBIR awards. Instead, grant recipients should be obtaining additional funds from a procurement program at the agency that made the award, from private investors,

---

[3]These are the 1982 Small Business Development Act, and the subsequent multiyear reauthorizations of the SBIR program in 1992 and 2000.

[4]With the agreement of the Small Business Administration, which plays an oversight role for the program, this amount can be higher in certain circumstances (e.g., drug development at NIH) and is often lower with smaller SBIR programs (e.g., EPA or the Department of Agriculture).

or from the capital markets. The objective of this phase is to move the technology from the prototype stage to the marketplace.

Obtaining Phase III support is often the most difficult challenge for new firms to overcome. In practice, agencies have developed different approaches to facilitate SBIR grantees' transition to commercial viability; not least among them are additional SBIR grants.

Previous NRC research has shown that firms have different objectives in applying to the program. Some want to demonstrate the potential of promising research but may not seek to commercialize it themselves. Others think they can fulfill agency research requirements more cost-effectively through the SBIR program than through the traditional procurement process. Still others seek a certification of quality (and the private investments that can come from such recognition) as they push science-based products towards commercialization.[5]

1.3 SBIR REAUTHORIZATIONS

The SBIR program approached reauthorization in 1992 amidst continued concerns about the U.S. economy's capacity to commercialize inventions. Finding that "U.S. technological performance is challenged less in the creation of new technologies than in their commercialization and adoption," the National Academy of Sciences at the time recommended an increase in SBIR funding as a means to improve the economy's ability to adopt and commercialize new technologies.[6]

Following this report, the Small Business Research and Development Enhancement Act (P.L. 102-564), which reauthorized the SBIR program until September 30, 2000, doubled the set-aside rate to 2.5 percent.[7] This increase in the percentage of R&D funds allocated to the program was accompanied by a stronger emphasis on encouraging the commercialization of SBIR-funded

---

[5]See Reid Cramer, "Patterns of Firm Participation in the Small Business Innovation Research Program in Southwestern and Mountain States," in National Research Council, *The Small Business Innovation Research Program, An Assessment of the Department of Defense Fast Track Initiative*, Charles W. Wessner, ed., Washington, D.C.: National Academy Press, 2000.

[6]See National Research Council, *The Government Role in Civilian Technology: Building a New Alliance*, Washington, D.C.: National Academy Press, 1992, p. 29.

[7]For fiscal year 2005, this has resulted in a program budget of approximately $1.85 billion across all federal agencies, with the Department of Defense having the largest SBIR program at $943 million, followed by the National Institutes of Health (NIH) at $562 million. The DoD SBIR program, is made up of ten participating components: Army, Navy, Air Force, Missile Defense Agency (MDA), Defense Advanced Research Projects Agency (DARPA), Chemical Biological Defense (CBD), Special Operations Command (SOCOM), Defense Threat Reduction Agency (DTRA), National Imagery and Mapping Agency (NIMA), and the Office of Secretary of Defense (OSD). NIH counts 23 separate institutes and agencies making SBIR awards, many with multiple programs.

technologies.[8] Legislative language explicitly highlighted commercial potential as a criterion for awarding SBIR grants. For Phase I awards, Congress directed program administrators to assess whether projects have "commercial potential," in addition to scientific and technical merit, when evaluating SBIR applications.

The 1992 legislation mandated that program administrators consider the existence of second-phase funding commitments from the private sector or other non-SBIR sources when judging Phase II applications. Evidence of third-phase follow-on commitments, along with other indicators of commercial potential, was also to be sought. Moreover, the 1992 reauthorization directed that a small business's record of commercialization be taken into account when evaluating its Phase II application.[9]

The Small Business Reauthorization Act of 2000 (P.L. 106-554) extended SBIR until September 30, 2008. It called for this assessment by the National Research Council of the broader impacts of the program, including those on employment, health, national security, and national competitiveness.[10]

1.4 STRUCTURE OF THE NRC STUDY

This NRC assessment of SBIR has been conducted in two phases. In the first phase, at the request of the agencies, a formal report on research methodology was to be developed by the NRC. Once developed, this methodology was then reviewed and approved by an independent National Academies panel of experts.[11] Information about the program was also gathered through interviews with SBIR program administrators and during four major conferences where SBIR officials were invited to describe program operations, challenges, and accomplishments.[12] These conferences highlighted the important differences in

---

[8]See Robert Archibald and David Finifter, "Evaluation of the Department of Defense Small Business Innovation Research Program and the Fast Track Initiative: A Balanced Approach," in National Research Council, *The Small Business Innovation Research Program: An Assessment of the Department of Defense Fast Track Initiative*, op. cit., pp. 211-250.

[9]A GAO report had found that agencies had not adopted a uniform method for weighing commercial potential in SBIR applications. See U.S. General Accounting Office, *Federal Research: Evaluations of Small Business Innovation Research Can Be Strengthened*, AO/RCED-99-114, Washington, D.C.: U.S. General Accounting Office, 1999.

[10]The current assessment is congruent with the Government Performance and Results Act (GPRA) of 1993: <http://govinfo.library.unt.edu/npr/library/misc/s20.html>. As characterized by the GAO, GPRA seeks to shift the focus of government decision-making and accountability away from a preoccupation with the activities that are undertaken—such as grants dispensed or inspections made—to a focus on the results of those activities. See <http://www.gao.gov/new.items/gpra/gpra.htm>.

[11]The SBIR methodology report is available on the Web. Access at <*http://www7.nationalacademies.org/sbir/SBIR_Methodology_Report.pdf*>.

[12]The opening conference on October 24, 2002, examined the program's diversity and assessment challenges. For a published report of this conference, see National Research Council, *SBIR: Program Diversity and Assessment Challenges*, Charles Wessner, ed., Washington, D.C.: The National Academies Press, 2004. A second conference, held on March 28, 2003, was titled, "Identifying Best

each agency's SBIR program's goals, practices, and evaluations. The conferences also explored the challenges of assessing such a diverse range of program objectives and practices using common metrics.

The second phase of the NRC study implemented the approved research methodology. The Committee deployed multiple survey instruments and its researchers conducted case studies of a wide profile of SBIR firms. The Committee then evaluated the results and developed both agency-specific and overall findings and recommendations for improving the effectiveness of the SBIR program. The final report includes complete assessments for each of the five agencies and an overview of the program as a whole.

1.5 SBIR ASSESSMENT CHALLENGES

At its outset, the NRC's SBIR study identified a series of assessment challenges that must be addressed. As discussed at the October 2002 conference that launched the study, the administrative flexibility found in the SBIR program makes it difficult to make cross-agency comparisons. Although each agency's SBIR program shares the common three-phase structure, the SBIR concept is interpreted uniquely at each agency. This flexibility is a positive attribute in that it permits each agency to adapt its SBIR program to the agency's particular mission, scale, and working culture. For example, NSF operates its SBIR program differently than DoD because "research" is often coupled with procurement of goods and services at DoD but rarely at NSF. Programmatic diversity means that each agency's SBIR activities must be understood in terms of their separate missions and operating procedures. This diversity is commendable but, operationally, makes the task of assessing the program more challenging.

A second challenge concerns the linear process of commercialization implied by the design of SBIR's three phase structure.[13] In the linear model, illustrated in Figure 1-1, innovation begins with basic research supplying a steady stream of fresh and new ideas. Among these ideas, those that show technical feasibility become innovations. Such innovations, when further developed by firms, become marketable products driving economic growth.

As NSF's Joseph Bordogna observed at the launch conference, innovation almost never takes place through a protracted linear progression from research to

---

Practice." The conference provided a forum for the SBIR Program Managers from each of the five agencies in the study's purview to describe their administrative innovations and best practices. A conference on June 14, 2005, focused on the commercialization of SBIR funded innovations at DoD and NASA. See National Research Council, *SBIR and the Phase III Challenge of Commercialization*, Charles W. Wessner, ed., Washington, D.C.: The National Academies Press, 2007. A final conference, held on April 7, 2006, examined role of the state programs in leveraging SBIR to advance local and regional economic growth.

[13]This view was echoed by Duncan Moore: "Innovation does not follow a linear model. It stops and starts." National Research Council, *SBIR: Program Diversity and Assessment Challenges*, op. cit.

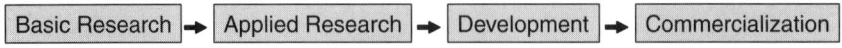

FIGURE 1-1 The linear model of innovation.

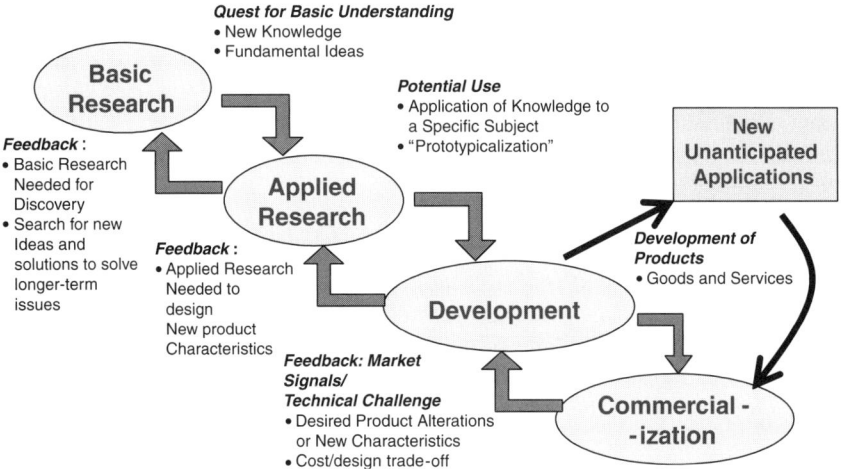

FIGURE 1-2 A feedback model of innovation.

development to market.[14] Research and development drives technological innovation, which, in turn, opens up new frontiers in R&D. True innovation, Bordogna noted, can spur the search for new knowledge and create the context in which the next generation of research identifies new frontiers. This nonlinearity, illustrated in Figure 1-2, makes it difficult to rate the efficiency of SBIR program. Inputs do not match up with outputs according to a simple function. Figure 1-2, while more complex than Figure 1-1 is itself a highly simplified model. For example, feedback loops can stretch backwards or forwards by more than one level.

A third assessment challenge relates to the measurement of outputs and outcomes. Program realities can and often do complicate the task of data gathering. In some cases, for example, SBIR recipients receive a Phase I award from one agency and a Phase II award from another. In other cases, multiple SBIR awards may have been used to help a particular technology become sufficiently mature to reach the market. Also complicating matters is the possibility that for any particular grantee, an SBIR award may be only one among other federal and non-federal sources of funding. Causality can thus be difficult, if not impossible, to establish. The task of measuring outcomes is made harder because compa-

---

[14]While few hold this process of linear innovation to be literally true, the concept nonetheless survives—for example, in retrospective accounts of the path taken by a particular innovation.

nies that have garnered SBIR awards can also merge, fail, or change their name before a product reaches the market. In addition, principal investigators or other key individuals can change firms, carrying their knowledge of an SBIR project with them. A technology developed using SBIR funds may eventually achieve commercial success at an entirely different company than that which received the initial SBIR award.

Complications plague even the apparently straightforward task of assessing commercial success. For example, research enabled by a particular SBIR award may take on commercial relevance in new unanticipated contexts. At the launch conference, Duncan Moore, former Associate Director of Technology at the White House Office of Science and Technology Policy (OSTP), cited the case of SBIR-funded research in gradient index optics that was initially considered a commercial failure when an anticipated market for its application did not emerge. Years later, however, products derived from the research turned out to be a major commercial success.[15] Today's apparent dead end can be a lead to a major achievement tomorrow. Lacking clairvoyance, analysts cannot anticipate or measure such potential SBIR benefits.

Gauging commercialization is also difficult when the product in question is destined for public procurement. The challenge is to develop a satisfactory measure of how useful an SBIR-funded innovation has been to an agency mission. A related challenge is determining how central (or even useful) SBIR awards have proved in developing a particular technology or product. In some cases, the Phase I award can meet the agency's need—completing the research with no further action required. In other cases, surrogate measures are often required. For example, one way of measuring commercialization success is to count the products developed using SBIR funds that are procured by an agency such as DoD. In practice, however, large procurements from major suppliers are typically easier to track than products from small suppliers such as SBIR firms. Moreover, successful development of a technology or product does not always translate into successful "uptake" by the procuring agency. Often, the absence of procurement may have little to do with the product's quality or the potential contribution of SBIR.

Understanding failure is equally challenging. By its very nature, an early-stage program such as SBIR should anticipate a high failure rate. The causes of failure are many. The most straightforward, of course, is *technical failure*, where the research objectives of the award are not achieved. In some cases, the project can be a technically successful but a commercial failure. This can occur when a procuring agency changes its mission objectives and hence its procurement priorities. NASA's new Mars Mission is one example of a *mission shift* that may result in the cancellation of programs involving SBIR awards to make room for

---

[15]Duncan Moore, "Turning Failure into Success," in National Research Council, *SBIR: Program Diversity and Assessment Challenges*, op. cit., p. 94.

new agency priorities. Cancelled weapons system programs at the Department of Defense can have similar effects. Technologies procured through SBIR may also *fail in the transition to acquisition.* Some technology developments by small businesses do not survive the long lead times created by complex testing and certification procedures required by the Department of Defense. Indeed, small firms encounter considerable difficulty in penetrating the "procurement thicket" that characterizes defense acquisition.[16] In addition to complex federal acquisition procedures, there are strong disincentives for high-profile projects to adopt untried technologies. Technology transfer in commercial markets can be equally difficult. A *failure to transfer to commercial markets* can occur even when a technology is technically successful if the market is smaller than anticipated, competing technologies emerge or are more competitive than expected, if the technology is not cost competitive, or if the product is not adequately marketed. Understanding and accepting the varied sources of project failure in the high-risk, high-reward environment of cutting-edge R&D is a challenge for analysts and policy makers alike.

This raises the issue concerning the standard on which SBIR programs should be evaluated. An assessment of SBIR must take into account the expected distribution of successes and failures in early-stage finance. As a point of comparison, Gail Cassell, Vice President for Scientific Affairs at Eli Lilly, has noted that only one in ten innovative products in the biotechnology industry will turn out to be a commercial success.[17] Similarly, venture capital funds often achieve considerable commercial success on only two or three out of twenty or more investments.[18]

In setting metrics for SBIR projects, therefore, it is important to have a realistic expectation of the success rate for competitive awards to small firms investing

---

[16]For a description of the challenges small businesses face in defense procurement, the subject of a June 14, 2005, NRC conference and one element of the congressionally requested assessment of SBIR, see National Research Council, *SBIR and the Phase III Challenge of Commercialization*, op. cit. Relatedly, see remarks by Kenneth Flamm on procurement barriers, including contracting overhead and small firm disadvantages in lobbying in National Research Council, *SBIR: Program Diversity and Assessment Challenges*, op. cit., pp. 63-67.

[17]Gail Cassell, "Setting Realistic Expectations for Success," Ibid, p. 86.

[18]See John H. Cochrane, "The Risk and Return of Venture Capital," *Journal of Financial Economics*, 75(1)3-52, 2005. Drawing on the VentureOne database Cochrane plots a histogram of net venture capital returns on investments that "shows an extraordinary skewness of returns. Most returns are modest, but there is a long right tail of extraordinary good returns. Fifteen percent of the firms that go public or are acquired give a return greater than 1,000 percent! It is also interesting how many modest returns there are. About 15 percent of returns are less than 0, and 35 percent are less than 100 percent. An IPO or acquisition is not a guarantee of a huge return. In fact, the modal or 'most probable' outcome is about a 25% return." See also Paul A. Gompers and Josh Lerner, "Risk and Reward in Private Equity Investments: The Challenge of Performance Assessment," *Journal of Private Equity*, 1(Winter 1977):5-12. Steven D. Carden and Olive Darragh, "A Halo for Angel Investors" *The McKinsey Quarterly*, 1, 2004, also show a similar skew in the distribution of returns for venture capital portfolios.

in promising but unproven technologies. Similarly, it is important to have some understanding of what can be reasonably expected—that is, what constitutes "success" for an SBIR award, and some understanding of the constraints and opportunities successful SBIR awardees face in bringing new products to market. From the management perspective, the rate of success also raises the question of appropriate expectations and desired levels of risk taking. A portfolio that always succeeds would not be investing in high risk, high pay-off projects that push the technology envelope. A very high rate of "success" would, thus, paradoxically suggest an inappropriate use of the program. Understanding the nature of success and the appropriate benchmarks for a program with this focus is therefore important to understanding the SBIR program and the approach of this study.

1.6 ASSESSING SBIR AT THE DEPARTMENT OF ENERGY (DOE)

In gathering and analyzing the data to assess the SBIR program at the Department of Energy, the Committee drew on the following set of research questions:

- How successful has DoE SBIR program been in **commercializing technologies** supported by Phase I and Phase II awards (and what are the factors that have contributed to or inhibited this level of commercialization)?
- To what extent has DoE SBIR program supported DoE's **mission** (and what are the factors that have contributed to or inhibited this level of support)?
- To what extent has DoE SBIR program **stimulated innovation**?
- How well has the DoE SBIR program encouraged **small firms** and supported the growth and development of woman- and minority-owned businesses?
- How effective has DoE's **management** of the SBIR program been (and how might this management be improved)?

1.6.1 Surveys of DoE SBIR Award-recipient Companies

Original data gathered by the research team in support of the NRC study of DoE SBIR program included a survey of DoE Phase II award-recipient firms; a survey DoE Phase I award-recipient firms that did not also receive a Phase II award; a survey of DoE technical staff involved in the SBIR program; numerous interviews with DoE personnel directly involved in administering the SBIR program; the assessment and analysis of data provided by DoE's SBIR staff; and ten company case studies.

The NRC Phase II Survey

In spring 2005, the NRC administered a survey of Phase II SBIR projects across agencies as part of its congressionally mandated evaluation of the SBIR

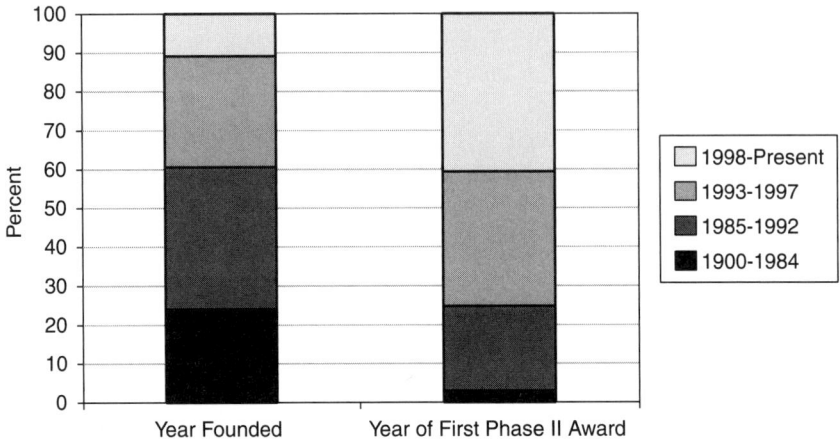

FIGURE 1-3 Project start dates for respondents in NRC study.

Program. The survey targeted a sample of Phase II awards that were awarded through 2001. A large majority of Phase II awards would have been completed by the 2005 survey date, and at least some commercialization efforts could have been initiated.

There may be some biases in these data. Projects from firms with multiple awards were underrepresented in the sample, because they could not be expected to complete a questionnaire for each of possibly numerous awards received; but they may have been overrepresented in the responses because they might be more committed to the SBIR program. Nearly 40 percent of respondents began Phase I efforts after 1998, partly because the number of Phase I awards increased, starting in the late 1990s, and partly because winners from more distant years are harder to reach, as small businesses regularly cease operations, staff with knowledge of SBIR awards leave, and institutional knowledge erodes.

For DoE, the sample size of Phase II projects targeted was 439. One hundred-fifty-seven respondents provided information on Phase II projects awarded by DoE, a response rate of approximately 36 percent. The response was considered adequate for drawing inferences.

The NRC Phase I Survey

The committee conducted a second recipient survey, in an attempt to determine the impact of Phase I awards that did not go on to Phase II. The original sample for this Phase I study was the 2,005 DoE Phase I awards from 1992-2001 inclusive. Valid responses were received from 155 DoE Phase I projects that did not advance to Phase II.

INTRODUCTION   21

Survey of DoE Project Managers

The technical project managers of individual SBIR projects can provide unique perspectives on the SBIR program. The project managers were surveyed electronically in three agencies—DoD, DoE, and NASA.

The Project Manager Survey was based on Phase II projects awarded during the study period (1992-2001 inclusive). Project managers for these projects were identified with the help of the agencies. As expected, there was significant attrition (because of the absence of email addresses, the inability to identify the project manager, the project manager having left the agency or becoming deceased, etc.). The three agencies were able to locate the names and email addresses of project managers for 2,584 projects. Of these, responses were received for 513 projects (a 20 percent response rate), of which 84 were for DoE projects (a 16 percent response rate). The number of individuals responding was fewer than the number of projects because some project managers had oversight for multiple projects.

1.6.2 Case Studies

Case studies can provide valuable insights concerning the viewpoints and concerns of the small businesses that participate in SBIR, insights that cannot be derived from statistical analysis. While all of the companies selected for case study won SBIR awards from the Department of Energy, most also won awards from other agencies as well. The interviews concerned their SBIR experience as a whole, and were not limited to DoE program.

Candidate case study firms were selected from four lists: top recipients of SBIR awards from the Department of Energy; DoE SBIR awardees who received R&D 100 awards; DoE identified "success stories;" and firms with large commercial sales as reported to DoE SBIR program. From a list of 34 candidate firms, 10 were selected, including firms from a variety of locations, across a range of founding dates, having received different numbers of SBIR awards received, and representing different technological domains.

The case study interviews focused on learning how the companies use the SBIR program: the extent to which SBIR is important to their company's survival and growth, whether and how they intend to commercialize SBIR technology, whether and how the receipt of multiple awards influence their ability to commercialize, what challenges they have faced in the commercialization process, in what way they see the SBIR program serving the needs of technology entrepreneurs and how they believe the program can be improved. In addition, we sought to learn how the companies were affected by the agencies' administration of the program and what suggestions the companies would have on how to improve program administration.

The case study companies range in age from 7 to 44 years old, in employees from 5 to 105, and in the number of SBIR awards from one to over a hundred. They cover seven states, both rural and urban areas, and present a variety of

Box 1-1
Multiple Sources of Bias in Survey Response

Large innovation surveys involve multiple sources of bias that can skew the results in both directions. Some common survey biases are noted below.[a]

- **Successful and more recently funded firms are more likely to respond.** Research by Link and Scott demonstrate that the probability of obtaining research project information by survey decreases for less recently funded projects and it increased the greater the award amount.[b] Nearly 40 percent of respondents in the NRC Phase II Survey began Phase I efforts after 1998, partly because the number of Phase I awards increased, starting in the mid 1990s, and partly because winners from more distant years are harder to reach. They are harder to reach as time goes on because small businesses regularly cease operations, are acquired, merge, or lose staff with knowledge of SBIR awards.
- **Success is self reported.** Self-reporting can be a source of bias, although the dimensions and direction of that bias are not necessarily clear. In any case, policy analysis has a long history of relying on self-reported performance measures to represent market-based performance measures. Participants in such retrospectively analyses are believed to be able to consider a broader set of allocation options, thus making the evaluation more realistic than data based on third party observation.[c] In short, company founders and/or principal investigators are in many cases simply the best source of information available.
- **Survey sampled projects at firms with multiple awards.** Projects from firms with multiple awards were under-represented in the sample, because they could not be expected to complete a questionnaire for each of dozens or more awards.
- **Failed firms are difficult to contact.** Survey experts point to an "asymmetry" in their ability to include failed firms for follow-up surveys in cases where the firms no longer exist.[d] It is worth noting that one cannot necessarily infer that the SBIR project failed; what is known is only that the firm no longer exists.
- **Not all successful projects are captured.** For similar reasons, the NRC Phase II Survey could not include ongoing results from successful projects in firms that merged or were acquired before and/or after commercialization of the project's technology. The survey also did not capture projects of firms that did not respond to the NRC invitation to participate in the assessment.
- **Some firms may not want to fully acknowledge SBIR contribution to project success.** Some firms may be unwilling to acknowledge that they received important benefits from participating in public programs for a variety of reasons. For example, some may understandably attribute success exclusively to their own efforts.
- **Commercialization lag.** While the NRC Phase II Survey broke new ground in data collection, the amount of sales made—and indeed the number of projects that generate sales—are inevitably undercounted in a snapshot survey taken at a single point in time. Based on successive data sets collected from NIH SBIR award recipients, it is estimated that total sales from all responding projects may be on the order of 50 percent greater than can be captured in a single survey.[e] This underscores the importance of follow-on research based on the now-established survey methodology.

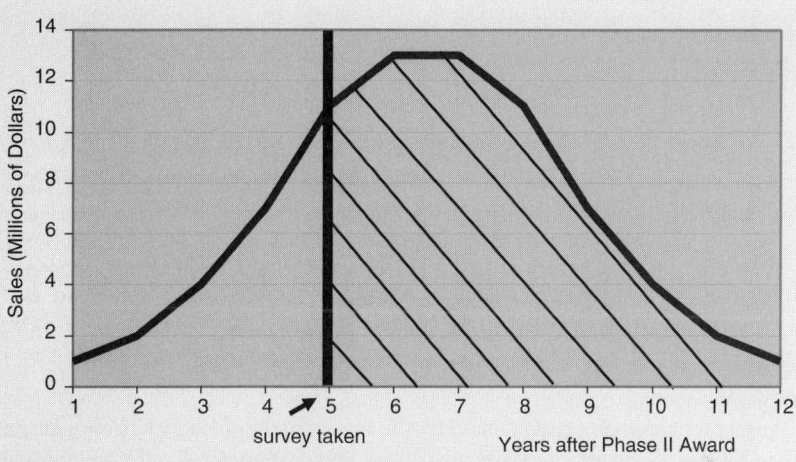

FIGURE 1-B-1 Survey bias due to commercialization lag.

These sources of bias provide a context for understanding the response rates to the NRC Phase I and Phase II Surveys conducted for this study. For the NRC Phase II Survey, of the 335 DoE firms that could be contacted out of a sample size of 439, 157 responded, representing a 47 percent response rate. The NRC Phase I Survey captured 14 percent of the 2,005 awards made by all five agencies over the period of 1992 to 2001. See Appendixes B and C for additional information on the surveys.

---

[a] For a technical explanation of the sample approaches and issues related to the NRC surveys, see Appendix B.
[b] Albert N. Link and John T. Scott, *Evaluating Public Research Institutions: The U.S. Advanced Technology Program's Intramural Research Initiative,* London: Routledge, 2005.
[c] While economic theory is formulated on what is called 'revealed preferences,' meaning individuals and firms reveal how they value scarce resources by how they allocate those resources within a market framework, quite often expressed preferences are a better source of information especially from an evaluation perspective. Strict adherence to a revealed preference paradigm could lead to misguided policy conclusions because the paradigm assumes that all policy choices are known and understood at the time that an individual or firm reveals its preferences and that all relevant markets for such preferences are operational. See {1} Gregory G. Dess and Donald W. Beard, "Dimensions of Organizational Task Environments," *Administrative Science Quarterly,* 29:52-73, 1984; and {2} Albert N. Link and John T. Scott, *Public Accountability: Evaluating Technology-Based Institutions,* Norwell, MA: Kluwer Academic Publishers, 1998.
[d] Albert N. Link and John T. Scott, *Evaluating Public Research Institutions: The U.S. Advanced Technology Program's Intramural Research Initiative,* op. cit.
[e] Data from NIH indicates that a subsequent survey taken two years later would reveal very substantial increases in both the percentage of firms reaching the market, and in the amount of sales per project. See National Research Council, *An Assessment of the Small Business Innovation Research Program at the National Institutes of Health,* Charles W. Wessner, ed., Washington, D.C.: The National Academies Press, 2008.

approaches to the SBIR program and to commercialization. The case study reports can be found in Appendix D.

1.7 STRUCTURE OF THE REPORT

The report is presented in eight chapters. Following this introduction, Chapter 2 lists the Committee assessment findings and recommendations for improving the operation of the SBIR program at the Department of Energy.

Chapter 3 provides some basic statistics concerning energy technology development and the SBIR program at DoE. The chapter begins with a brief summary of national trends in the funding of energy research and development before presenting SBIR Phase I and Phase II pertaining to the number of awards granted, award size, and geographical location of award-recipient firms over the period of study.

Chapter 4 examines actions taken by the Department of Energy to encourage commercialization efforts by SBIR awardees; how commercialization outcomes are measured; and what the various measures described indicate with respect to the commercialization of SBIR-supported technologies.

Chapter 5 focuses on the manner in which the program supports the mission of the Department of Energy "to advance the national, economic, and energy security of the United States; to promote scientific and technological innovation in support of that mission; and to ensure the environmental cleanup of the national nuclear weapons complex." The section highlights the tension that exists between commercialization and agency mission objectives of the program.

Chapter 6 describes the DoE program's support of woman- and minority-owned technology companies.

Chapter 7 addresses the manner in which the program spurs knowledge creation and the development of new technologies, including the strengthening of knowledge networks involving small business, and the creation of codified knowledge in the form of patents and publications.

Finally, Chapter 8 assesses the management of the program, focusing on program: outreach, the application process, award management, and program structure. It describes some developments in the administration of the program that took place following the study period, but prior to the completion of this report.

2

Findings and Recommendations

I. NRC STUDY FINDINGS

The Department of Energy (DoE) SBIR program is making significant progress in achieving the congressional goals for the program. The SBIR program is sound in concept and effective in practice at DoE. With the programmatic changes recommended here, the SBIR program should be even more effective in achieving its legislative goals.[1]

A. **Overall, the program has made significant progress in achieving its congressional objectives by:**

- Stimulating technological innovation; (see Finding G)
- Using small business to meet federal research and development needs; (see Finding C and D)
- Fostering and encourage participation by minority and disadvantaged persons in technological innovation; (see Finding E) and
- Increasing private sector commercialization of innovations derived from federal research and development. (See Finding B.)

---

[1] These objectives are set out in the Small Business Innovation Development Act (PL 97-219). In reauthorizing the program in 1992, (PL 102-564) Congress expanded the purposes to "emphasize the program's goal of increasing private sector commercialization developed through Federal research and development and to improve the Federal government's dissemination of information concerning small business innovation, particularly with regard to woman-owned business concerns and by socially and economically disadvantaged small business concerns."

B. **The DoE SBIR program is focused on commercialization and has seen meaningful achievement. There are, nonetheless, opportunities for improvement in commercialization.**

1. **A significant percentage of DoE SBIR projects are commercialized to some degree.**
    i. **Reaching the market.** NRC Phase II Survey data indicate that 41 percent of SBIR-funded projects reach the marketplace or have commercialization underway.[2] Using a different methodology, DoE's internal survey reported that 30.3 percent of firms reached the market.
    ii. **Revenue skew.** The NRC Phase II Survey data also show that a much smaller number (4 percent) of projects generate more than $5 million in revenues.[3] The distribution of sales resulting from SBIR awards from DoE is not *qualitatively* different from the distribution of returns from private sector investments.[4] These few projects have, however, been extremely successful; in the case of Atlantia Offshore, for example, a single Phase I and Phase II pair of awards resulted directly in a product that generated over $500 million in sales, in addition to other societal benefits. DoE's survey of their Phase II awardees indicated that 11 percent of companies reported sales greater than $850,000.
    iii. **Licensing revenue.** The NRC Phase II Survey indicates that licensing revenues have not been a significant source of additional commercial success at DoE.[5]

---

[2] See Figure 4-1. Twenty-four percent of NRC Phase II Survey respondents reported products/services/or processes in use at the time of the survey, and 18 percent reported commercialization underway (figures rounded). See NRC Phase II Survey, question 1.

[3] See NRC Phase II Survey, Question 4b.

[4] As with investments by angel investors or venture capitalists, SBIR awards result in highly concentrated sales, with a few awards accounting for a very large share of the overall sales generated by the program. These are appropriate referent groups, though not an appropriate group for direct comparison, not least because SBIR awards often occur earlier in the technology development cycle than where venture funds normally invest. Nonetheless, returns on venture funding tend to show the same high skew that characterizes commercial returns on the SBIR awards. See John H. Cochrane, "The Risk and Return of Venture Capital," *Journal of Financial Economics*, 75(1):3-52, 2005. Drawing on the VentureOne database, Cochrane plots a histogram of net venture capital returns on investments that "shows an extraordinary skewness of returns. Most returns are modest, but there is a long right tail of extraordinary good returns. Fifteen percent of the firms that go public or are acquired give a return greater than 1,000 percent! It is also interesting how many modest returns there are. About 15 percent of returns are less than 0, and 35 percent are less than 100 percent. An IPO or acquisition is not a guarantee of a huge return. In fact, the modal or "most probable" outcome is about a 25 percent return." See also Paul A. Gompers and Josh Lerner, "Risk and Reward in Private Equity Investments: The Challenge of Performance Assessment," *Journal of Private Equity,* 1(Winter):5-12, 1977. Steven D. Carden and Olive Darragh, "A Halo for Angel Investors," *The McKinsey Quarterly,* 1, 2004, also show a similar skew in the distribution of returns for venture capital portfolios.

[5] NRC Phase II Survey, Question 3.

FINDINGS AND RECOMMENDATIONS 27

    iv. **DoE awards are characterized by a relatively high degree of private sales**. Of the DoE SBIR award recipient firms that responded to the NRC Phase II Survey and reported sales of some type, 76 percent sold to domestic private sector firms and 14 percent to export markets.[6]

    v. **Limits on potential for broader commercialization**. In cases where the market is inherently limited, as with the case with sensitive energy technologies, products developed with DoE SBIR assistance often cannot become a large commercial successes.[7]

2. **DoE has a substantial history of commercialization support and has expanded its efforts in recent years**.

    i. DoE has sponsored a *Commercialization Assistance Program* (CAP) for the past 17 years. That program has provided, on a voluntary basis to Phase II awardees, individual assistance in developing business plans and in preparation of presentations to potential investment sponsors. It is operated by a contractor, Dawnbreaker, Inc., a private firm based in Rochester, NY.

    ii. For the past 6 years, DoE has offered their Phase II awardees additional market identification and networking service which requires much less time commitment from awardees than the very intensive CAP.[8] These programs have provided professional assistance in business plan development and/or market evaluation to participating SBIR companies.

    iii. Systematic tracking is necessary to assess the impact of these efforts on commercialization.[9]

3. **Third-party investors can be encouraged by the validation effect of SBIR funding**.

    i. **Additional investments**. Sixty-three percent of respondents indicated that they had received or made additional investments in the surveyed project, averaging just under $1 million per project.[10]

    ii. **No venture funding**. For a number of reasons (at least until recently) the energy sector in general and therefore DoE SBIR projects have not been attractive to venture capitalists. No responding projects indicated that they had received venture capital funding.[11]

---

[6]See Table 4-2.
[7]The case study of Diversified Technologies illustrates this phenomenon with regard to specialized transformers. See Appendix D.
[8]DoE SBIR publications and Web site; interviews with DoE SBIR staff.
[9]DoE calls for its contractors, Dawnbreaker and Foresight, to track their clients' performance for at least two years.
[10]See NRC Phase II Survey, Questions 22 and 23.
[11]See Table 4-4.

TABLE 2-1 2005 DoE SBIR Initiatives in Support of Commercialization

	CAP	Trailblazer	Technology Niche Assessment (TNA)	Virtual Deal Simulator™ (VDS)
Contractor	Dawnbreaker, Inc.	Foresight Science and Technology, Inc.	Foresight Science and Technology, Inc.	Foresight Science and Technology, Inc.
Start Date	January 2005	January 2005	January 2005	January 2005
Completion Date	December 2007	December 2007	December 2007	December 2007
Eligibility	SBIR II only	SBIR I only	SBIR I or II	SBIR I or II

SOURCE: Department of Energy SBIR program publications and Web site.

iii. **Acquisition**. In some cases, the technology developed had sufficient commercial potential that investors bought the grantee company outright.

iv. **Multiple other funding sources**. Many grantees have found additional funds from a wide range of sources, including angel funding. Sixty-three percent of NRC survey respondents attracted some additional investment (excluding further SBIR awards).

4. **Commercialization and selection**. Commercialization potential is now formally recognized as a selection criterion at DoE; it accounts for one-sixth of the total score used in selecting applications for award.[12]

C. **The DoE SBIR program is operated in alignment with the department's mission. It has the potential to contribute to the missions of the National Laboratories.**

1. **Effective mission alignment**. All DoE awards appear to be selected primarily on the basis of their potential to advance knowledge and provide solutions in the field of energy. There is no evidence that DoE awards are made in fields outside those linked to the department's mission.[13]

---

[12]Interviews with DoE SBIR staff.

[13]Possibly the most dramatic identified success from the program has been Atlantia's SeaStar® technology, a cost-effective alternative for the development of otherwise inaccessible small oil fields in deep water. The technology generated more than $500 million in revenues for Atlantia.

FINDINGS AND RECOMMENDATIONS *29*

2. **Agency technical managers are deeply involved in topic development and selection.**
   i. The impact of an SBIR project on DoE's mission to advance economic and energy security is carefully considered during the selection process. Awardees and DoE staff note that impact effects are an important component in every application. In all the cases examined, DoE SBIR funded projects related to the science and technology of energy.
   ii. Results from the NRC survey of Technical Topic Managers indicate that of the DoE projects surveyed, 70 percent reported that the project manager was involved in the generation of the topic that led to SBIR award. Fifty-eight percent of project managers reported involvement with the technology before Phase I began.
3. **Limited participation of DoE National Laboratories in SBIR.**
   i. While the National Laboratories are an important source of technical reviewers for proposals (which is an important component of the administration of the SBIR program at DoE), the Laboratories themselves are not otherwise strongly involved in the SBIR program.
   ii. As a result, the potentially significant role of the National Laboratories as partners with SBIR-award-recipient firms is not fully realized.

D. **The SBIR program at DoE has provided significant support for small business, frequently acting as the impetus for project deployment and the foundation of new firms.**

   1. **SBIR awards from the Department of Energy fund the development of technologies that, otherwise, might have developed more slowly, if at all.**
      i. **The project initiation decision.** More than 80 percent of NRC Phase II Survey respondents reported that they would likely or certainly not have gone forward with their project in the absence of SBIR funding.[14]
      ii. **New firm formation.** Twenty-three percent of DoE respondents indicated that their companies were founded entirely or partly because of an SBIR award. Of the NRC Phase I Survey respondents, 11 percent stated that their firms were founded or remained in businesses due to SBIR Phase I funding they were awarded. Growth within companies also occurred, with 34 percent of responding

---

[14]See Figure 7-1.

Phase I companies hiring one or more new employees as a direct result of an award.[15]

iii. **Critical source of early funding.** Case study work suggests that alternative very early stage venture funding for SBIR projects is nearly nonexistent. Without exception, all of the case study companies indicated that SBIR was vital to the development of their technology. Most suggested that the technology would not have been created if there had been no SBIR program. All credited SBIR as having played a significant role in the company's formulation or development.

2. **The DoE SBIR program supports the engagement of small business in federal R&D.**
   i. **Contracting mechanism.** SBIR provides a contracting mechanism that keeps small firms engaged in the innovation system. Many small firms find other approaches to seeking federal or corporate R&D support to be either too costly from a contracting standpoint or require too great a loss of control to be viable. This appears to be one reason why SBIR continues to be enthusiastically supported by the small business community despite the fact that the award sizes have stayed constant for over a decade.
   ii. **Application process as a filter.** At the same time, the application process—though simpler than that for other federal R&D programs—is nonetheless very demanding for a small firm, and the success rate of Phase I applications is typically at 18-20 percent.[16]

3. **The SBIR program also provides additional benefits for participating small businesses.**
   i. **Networks and partnering.** SBIR facilitates technological innovation through the creation of new partnerships and the strengthening of networks of innovators. Among case study firms, networking was identified as critical when attempting to find partners to assist with commercialization.[17]
   ii. **Alternative path development.** Companies often use SBIR to fund alternative development strategies, exploring technological options in parallel with other activities.
   iii. **Expand the company's technology base.** SBIR awards can help build a company's technology base, allowing the company to pursue

---

[15]Case studies of NanoSonic and Thunderhead Engineering, and Diversified Technologies, Inc.

[16]DoE SBIR program data prepared for the Subcommittee on Technology and Innovation, Committee on Science and Technology, U.S. House of Representatives, June 26, 2007. This success rate is subject to some variance—as high as 25 percent and as low as 12.5 percent in some years.

[17]See Figure 7-2.

FINDINGS AND RECOMMENDATIONS

an expanded suite of research projects, bring in researchers with new points of view and knowledge, and develop new markets.[18]

iv. **Stimulate other forms of financing**. SBIR awards can help a firm attract other forms of financing, such as funding opportunities from states (NexTech). Banks may also be more willing to provide business loans to companies that have received a Phase II grant, recognizing that a Phase II award is a good indicator of an innovation's potential marketability.[19]

v. **Enhance credibility with federal agencies**. SBIR awards typically lend credibility to a firm and enable it to earn other federal grants and contracts, including those obtained through Broad Area Announcements.[20]

vi. **Enhance credibility with private-sector firms**. SBIR awards can provide a "stamp of approval," allowing smaller companies to better access large private sector companies, including government contractors, and attract private investors. An SBIR award can instill confidence that the smaller firm, and its product, have been vetted to some degree, thus raising interest. Once one large company becomes aware of an SBIR firm, its reputation grows and others are likely to hear about the firm, its product, and its area of expertise.[21]

E. **Support for minority- and woman-owned firms.**

1. **No long-term trend is apparent**. Data from SBA indicates that between 1992 and 2005, the Phase I share of woman- and minority-owned businesses at DoE averaged just over 20 percent, with a decline in the early years of this decade, followed by an increase, accelerating in 2005 (see Figure 2-1). Data for Phase II are similar, although slightly lower. On average, woman- and minority-owned firms won 22.1 percent of Phase I awards from 2001-2005, and 19.1 percent of Phase II awards.[22]

2. **Lagging application success rates**. For Phase I, applications from woman-owned businesses have had a lower rate of success compared to all other groups—by approximately 3-10 percentage points—in

---

[18]See case studies of NanoScience, PPL, Diversified Technologies, Inc., and Thunderhead Engineering in Appendix D. See also the case study of Pearson Knowledge Technologies in National Research Council, *An Assessment of the Small Business Innovation Research Program at the Department of Defense*, Washington, D.C.: The National Academies Press, 2007 Prepublication.
[19]See the Airak case study in Appendix D.
[20]See the IPIX and NexTech case studies in Appendix D.
[21]See the case studies of Atlantia, NanoScience, and NexTech in Appendix D.
[22]See Chapter 6.

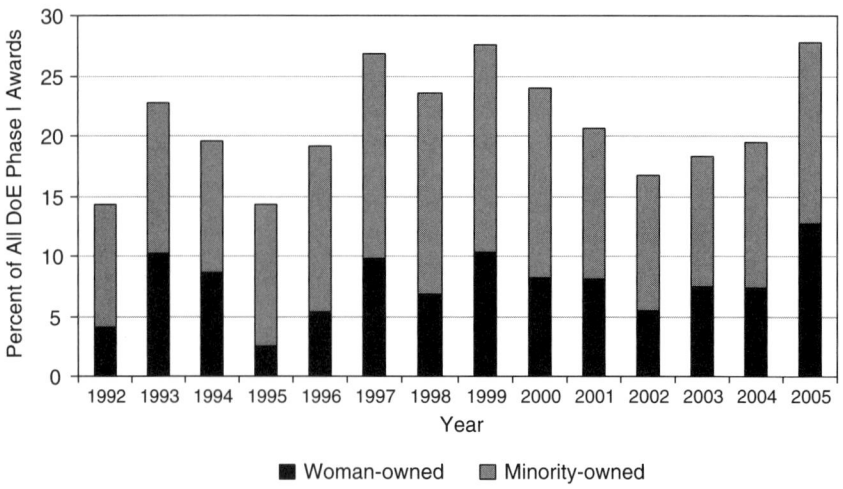

FIGURE 2-1 Shares of Phase I awards for woman- and minority-owned businesses, 1992-2005.
SOURCE: U.S. Small Business Administration, Tech-Net Database.

every year except one. For minority-owned companies, the success rate is better than for woman-owned companies but still lags behind the "other" category (neither woman-owned nor minority-owned). For 2002-2003, the success rate of minority-owned businesses was considerably lower than that for woman-owned and all other businesses.[23]

F. Understanding firms winning multiple awards.

1. **Wide distribution of funding.**
   i. SBA awards data indicates that between 1992 and 2005, 3,698 Phase I awards went to 1,459 different companies. The top 20 winners accumulated 556 Phase I awards, or 8 percent of all awards made during this period. No firm received 50 or more DoE Phase I awards.
   ii. Only five companies have been identified as receiving more than 15 Phase II awards between FY1992 and FY2003, and only two received more than 20, with the maximum being 21.[24]

---

[23]Ibid.

[24]The top 20 percent of winning companies together received 11.1 percent of awards. This is significantly lower than the Department of Defense.

FINDINGS AND RECOMMENDATIONS

2. Some products, involving complex, multidisciplinary technologies, require multiple awards and a long gestation to develop.
  i. Products developed within by SBIR-awardee firms often require multiple SBIR awards and/or other sources of funding. Many firms work in complex technology areas, where capabilities in complementary technologies are the basis for competitive success.
  ii. Developing prototypes of such products, with associated business plans and reliably estimated costs of production, can take a decade or more to complete.

3. Some SBIR firms with multiple awards achieve commercial impacts via spin-off firms.
  i. Some DoE-SBIR firms with multiple awards specialize in solving practical problems posed by government. The approach of Creare, Inc. in Hanover, New Hampshire, for example, is to create and transfer to spin-offs technologies that have particular commercial potential. As a consequence, Creare alone has spawned 15 spin-off firms employing over 1,500 people, with annual revenues reportedly in excess of $250 million.[25]

G. Knowledge creation and dissemination effects.

1. The SBIR program at DoE supports knowledge transfer in several ways.
  i. Patents. SBIR companies at DoE have generated numerous patents and publications, the traditional measures of knowledge transfer activity.[26] Forty-three percent of the projects responding to the NRC survey reported at least one patent application. Of these, 37 percent reported receiving a patent related to the SBIR project.
  ii. Peer-reviewed publications. Forty-six percent of projects surveyed by NRC resulted in at least one peer-reviewed article.[27]
  iii. Knowledge transfer from universities. The NRC survey also suggests that SBIR awards are supporting the transfer of knowledge, firm creation, and partnerships between universities and the private sector:[28]

---

[25] As a point of comparison, Xerox Technologies Ventures, the famed venture capital arm of Xerox Corporation, generated 35 spin-off firms over a comparable time period.

[26] See NRC Phase II Survey, Question 18.

[27] Without detailed identifying data on these patents and publications, it is not feasible to apply bibliometric and patent analysis techniques to assess the relative importance of these patents and publications.

[28] See Table 4-14 in National Research Council, *An Assessment of the Small Business Innovation Research Program*, Charles W. Wessner, ed., Washington, D.C.: The National Academies Press, 2008.

- o In more than two-thirds of responding companies at all agencies, at least one founder was previously an academic;
- o About one-third of company founders were most recently employed as academics before the creation of their company;
- o At DoE, about 34 percent of projects had some alignment with a university, through the use of university faculty as contractors on the project, use of universities as subcontractors, or employment of graduate students.

iv. **Indirect paths**. There is anecdotal evidence concerning beneficial "indirect path" effects—that projects provide investigators and research staff with knowledge that may later become relevant in a different context—often in another project or even another company. While these effects are not directly measurable, discussion during interviews and case studies suggest they exist.[29]

2. **SBIR research quality**.
   i. The NRC survey of technical topic managers (TTM's) indicated that technical managers see research funded by SBIR as being largely similar in quality to the research funded under other programs.
   ii. Another survey question asked whether more high quality research proposals were received than could be funded. Just under two-thirds (62 percent) of DoE project managers reported more fundable projects than were funded.[30] Success rates for SBIR Phase I competitions are consistently below 20 percent. In most years, the success rates have been about 15 percent.

H. **The DoE SBIR program has not benefited from regular evaluation.**

1. **Prior to the congressional legislation authorizing this study, no systematic, external program assessment had been undertaken at DoE. As a result, program management is not sufficiently evidence-based.**
   i. Partly as a result of insufficient resources, the program has insufficient data collection, limited reporting requirements, and limited analytic functions. This limits the program's capacity for self-assessment and adjustment.

---

[29]For a discussion of the "indirect path" phenomenon with regard to the results of innovation awards, see Rosalie Ruegg, "Taking a Step Back: An Early Results Overview of Fifty ATP Awards," in National Research Council, *The Advanced Technology Program: Assessing Outcomes,* Charles W. Wessner, ed., Washington, D.C.: National Academy Press, 2001.

[30]Thirty-one percent reported about the right number of proposals, and eight percent reported fewer fundable proposals than funds available.

ii. This lack of assessment means that program management does not have adequate ongoing information about how their actions affect outcomes such as commercialization, knowledge generation, and networking.

2. **Developing a culture of assessment.** DoE has recently developed a system for tracking outcomes internally. This represents a positive step towards an assessment culture, but a range of issues still need to be addressed and a more systematic approach to evaluation adopted.

I. **Limited resources for program management.**

1. **The fraction of resources devoted to SBIR program management at DoE is the lowest among agencies studied by the NRC.**
    i. Given the size of the program, the number of applicants each year, and the requirements of the award process, the SBIR staff at DoE is the minimum required to operate the program.
    ii. DoE SBIR staff devote nearly all their time to managing the processes for generating technical topics, and for receiving, evaluating, and selecting grant applications. This leaves little time for other important tasks such as outreach, measuring Phase III activity, encouraging Phase III activity (both within and outside the department, including the national laboratories), documenting successes, understanding failure, and developing and implementing program improvements.

2. **With most staff time devoted to compliance, there is limited opportunity for program enhancement.**
    i. **Impact on assessment.** Program managers do not have time or sufficient resources to support interaction between SBIR firms and technical staff; conduct internal and external assessment of the performance of funded firms; and evaluate the commercialization program and the overall effectiveness of awards.
        o **Limited post-award follow up.** While DoE's technical staff have taken the lead in developing topics for the program solicitation, they have had very little interaction with the companies that receive SBIR awards.[31]
        o **Limited monitoring and assessment.** Administrative staff or technical staff report few opportunities to visit funded firms or otherwise track their progress. Over the course of this NRC

---

[31] Interviews with DoE SBIR staff. Progress is being made in this regard. DoE technical staff has recently demonstrated its direct involvement in a newly instituted Continuation Application Process.

study, the DoE SBIR staff has sought new approaches to encourage DoE's technical staff to become familiar with the work of SBIR funded companies.
o **Limited evaluation**. While the DoE SBIR staff maintain a detailed database of commercialization outcomes reported by funded firms, they do not as yet have the resources to validate the reported data or employ them in formal program evaluation. Evaluation of the effectiveness of commercialization programs has also been lacking. Program staff do not have adequate resources to assess the relative effectiveness of different approaches undertaken by DoE to enhance commercialization outcomes.

ii. **Impact on cycle time**. DoE provides only one solicitation annually. This contrasts in particular with DoD and NIH, which offer candidates several opportunities to propose projects. However, multiple solicitations require more resources, which are currently not available to the at DoE SBIR program.

iii. **Impact on gaps between SBIR Phase I and Phase II funding**. The funding gap between Phase I and Phase II also affects companies. More than half the respondents to the NRC Phase II Survey reported that they stopped work on their project during this period. A small number of respondents (3 percent) reported ceasing operations entirely in this unfunded interval. DoE has not adopted any of the measures to reduce the Phase I-Phase II gap implemented at other agencies.[32]

iv. **Impact on outreach**. Citing time and resource constraints, DoE SBIR staff decline most invitations to speak about the program. The agency does participate in the SBIR National Conferences, sponsored by DoD and NSF. However, DoE has attempted to avoid state and local conferences, largely because the limited resources and available staff are focused on day-to-day program operations.[33]

II. NRC STUDY RECOMMENDATIONS

The recommendations in this section are designed to improve the operation of the SBIR program at the Department of Energy. They complement the core findings that the program is addressing its legislative goals—that significant commercialization is occurring, that the awards are making valuable additions to nation's stock of scientific and technical knowledge, and that SBIR is developing products that apply this knowledge to the Department of Energy's missions.

---

[32]See NRC Phase II Survey, Question 28.
[33]Interviews with DoE SBIR program managers.

A. **Develop linkages between DoE technical staff and SBIR awardees.**
  1. **Provide contact information.** Communication with applicants, particularly first-time applicants, is critical to the processes. DoE should increase the opportunity for all potential applicants to interact with the technical program managers, including the publication of their names and contact information within the annual solicitation.[34]
  2. **Provide pre-release information on topics.** Pre-proposal communication through a fair, transparent mechanism is likely to improve proposal quality and, therefore, overall program effectiveness. Consider the DoD Pre-Release information exchange model, whereby the relevant technical officer for each topic is available via email and phone for questions during a period before the official release of the solicitation. At DoD this practice has had the effect of ensuring not only that applicants have a better understanding of the work being solicited, but also in generating higher quality and better focused proposals.

B. **Improve commercialization support.**
  1. **Develop a match-making function for SBIR awardees.** Bring SBIR participants together with potential corporate customers, perhaps in trade show, technical challenge workshop, or technology demonstration/validation formats. These functions could include large corporations identified by the agency's two SBIR commercialization assistance vendors, including large energy technology corporations that serve as DoE contractors.
  2. **A possible model.** The Navy Opportunity Forum, initially based on a DoE initiative, is well funded. Similar levels of funding and scale might enhance DoE's return on this type of activity. Another model is the National Renewable Energy Laboratory's Enterprise Growth Forum.

C. **Explore reallocation of funding and topics between programs.**
  1. While SBIR funds are awarded by DoE using a competitive review of the best submitted proposals, the success rates of Phase I awards vary dramatically between the different programs. This disparity occurs because DoE allocates SBIR awards back to individual programs based upon a running average of the amount of SBIR funds contributed by that program, rather than only on proposal quality.
  2. DoE might consider whether current adjustments on an ad hoc basis are sufficient, or whether a policy that allows individual DoE programs to benefit from SBIR projects in excess of their contributed amounts would prove more effective.

---

[34]DoE initiated an information release process in FY2006. The relevant technical officer is available via email from the opening date to the closing date of the Funding Opportunity Announcement.

3. By allowing DoE programs to compete among themselves for a greater share of total, SBIR can better address the agency's mission and encourage proposals of the highest quality.

D. Engage the National Laboratories: More outreach to the National Laboratories is desirable. This could include:
1. **Subcontracting**. Encourage participation of National Laboratories as subcontractors to the small business on SBIR projects, not least by removing regulatory barriers to the use of National Laboratories as subcontractors. More resources are likely to be required to carry out these activities.
2. **Improve tracking**. Develop procedures to track the relationship between National Laboratories and the SBIR program more formally, including the documentation of Phase III successes.

E. Increase the participation and success rates of woman- and minority-owned firms in the SBIR program.
1. **Encourage participation**. Develop targeted outreach to improve the participation rates of woman- and minority-owned firms, and strategies to improve their success rates. These outreach efforts and other strategies should be based on an analysis of past proposals and feedback from the affected groups.[35]
2. **Improve data collection and analysis**.
    i. There appears to be room for further improvement in the participation of women and minorities in DoE SBIR program. DoE should undertake efforts to assess why woman- and minority-owned companies have experienced relatively lower success rates and to examine courses of action that may rectify this underrepresentation.
    ii. The Committee also strongly encourages DoE to gather the data that would track woman and minority firms as well as principal investigators (PIs), and to ensure that SBIR is an effective road to opportunity. The success rates of woman and minority Principle Investigators and Co-Investigators can also provide a measure of woman and minority participation in the SBIR program.
3. **Encourage emerging talent**. The number of women and, to a lesser extent, minorities graduating with advanced scientific and engineering degrees has been increasing over the past decade. This means that many of the woman and minority scientists and engineers with the advanced degrees usually necessary to compete effectively in the SBIR

---

[35]This recommendation should not be interpreted as lowering the bar for the acceptance of proposals from woman- and minority-owned companies, but rather as assisting them to become able to meet published criteria for grants at rates similar to other companies on the basis of merit, and to ensure that there are no negative evaluation factors in the review process that are biased against these groups.

program are relatively young and may not yet have arrived at the point in their careers where they own their own companies. However, they may well be ready to serve as principal investigators (PIs) and/or senior co-investigators (Co-Is) on SBIR projects. Over time, this talent pool could become a promising source of SBIR participants.[36]

F. **Develop data for evaluation and conduct regular assessments.**
1. **Summary annual report.** DoE should annually provide Congress with an enhanced summary report on the SBIR program. This should include descriptive statistics for applications, awards, and outcomes along the dimensions identified in this report, including knowledge creation, technology innovation, and impact on agency mission, as well as commercialization. As part of this process, DoE should produce regular reports from the commercialization database.
2. **Company Commercialization Report.** DoE may also wish to consider whether the DoD model—and technology—for producing a Company Commercialization Report, updated each time the company applies for further awards, might be a useful way of generating better data about commercial outcomes.
3. **Regular assessments**. The proposed annual report noted above could become a focus for wider efforts to develop improved internal assessment capabilities that can be used to enhance program operations. It could also tie proposed improvements to data-driven analysis and, for example, include an evaluation of the predictive power of selection scoring with regard to commercialization and other outcomes. DoE should also commission regular external arms-length evaluations to assess the program progress and the impact of new initiative.

G. **Provide additional management funding to develop and maintain a results-oriented program with a focused evaluation culture**.
1. Effective oversight relies on appropriate funding. A data-driven program requires high quality data and systematic assessment. As noted above, sufficient resources are not currently available for these functions.
2. Increased funding is needed to provide effective oversight, including site visits, program review, systematic third-party assessments, and other necessary management activities.

---

[36]Academics represent an important future pool of applicants, firm founders, principal investigators, and consultants. Recent research shows that owing to the low number of women in senior research positions in many leading academic science departments, few women have the chance to lead a spin-out. "Underrepresentation of female academic staff in science research is the dominant (but not the only) factor to explain low entrepreneurial rates amongst female scientists." See Peter Rosa and Alison Dawson, "Gender and the commercialization of university science: academic founders of spinout companies," *Entrepreneurship & Regional Development*, 18(4):341-366, 2006.

3. To enhance program utilization, management, and evaluation, additional funds should be provided. There are three ways that this might be achieved:
    i. Additional funds might be allocated internally, within the existing budgets of the services and agencies, as the Navy has done, with commensurately positive results.
    ii. Funds might be drawn from the existing set-aside for the program to carry out these activities.
    iii. The set-aside for the program, currently at 2.5 percent of external research budgets, might be increased slightly, with the goal of providing additional resources to maximize the program's return to the nation.[37]
4. These recommended improvements should enable the DoE SBIR managers to address the four mandated congressional objectives in a more efficient and effective manner.

H. **DoE should consider the creation of an independent advisory board that draws together senior agency management, SBIR managers, and other stakeholders as well as outside experts to review current operations and achievements and recommend changes to the SBIR program.**
1. Augmenting the role of the current DoE SBIR Oversight Committee, the purpose of such an advisory board is to provide a regular monitoring and feedback mechanism that would address the need for upper management attention, and encourage internal evaluation and regular assessment of progress towards definable metrics.
2. The annual report of the DoE SBIR program, recommended above, could be presented to the board. The board would review the report that would include updates on program progress, management practices, and make recommendations to senior agency officials.

---

[37]Each of these options has its advantages and disadvantages. For the most part, the departments, institutes, and agencies responsible for the SBIR program have not proved willing or able to make additional management funds available. Without direction from Congress, they are unlikely to do so. With regard to drawing funds from the program for evaluation and management, current legislation does not permit this and would have to be modified, therefore the Congress has clearly intended program funds to be for awards only. The third option, involving a modest increase to the program, would also require legislative action and would perhaps be more easily achievable in the event of an overall increase in the program. In any case, the Committee envisages an increase of the "set aside" of perhaps 0.03 percent to 0.05 percent on the order of $35 million to $40 million per year or, roughly, double what the Navy currently makes available to manage and augment its program. In the latter case (0.05 percent), this would bring the program "set aside" to 2.55 percent, providing modest resources to assess and manage a program that is approaching an annual spend of some $2 billion. Whatever modality adopted by the Congress, without additional resources the Committee's call for improved management, data collection, experimentation, and evaluation may prove moot.

3. The board could be assembled on the model of the Defense Science Board (DSB) or perhaps the National Science Foundation's Advisory Board.[38] In any case, it should include senior agency staff and the Director's Office on an *ex officio* basis, and bring together, *inter alia*, representatives from industry (including award recipients), academics, and other experts in program management.

---

[38]The intent here is to use the DSB or the NSF Board as a model, not something necessarily to be copied exactly.

3

Award Statistics

The Department of Energy (DoE) administers the federal government's third largest SBIR program, with an annual budget of approximately $104 million in FY2005. Most nuclear weapons programs, a major part of DoE's overall R&D activities, are however excluded from SBIR.[1]

In this chapter, we review some overall statistics about DoE's Small Business Innovation Research (SBIR) program. In particular, we examine the number and amount of Phase I and Phase II awards, and break down the data in terms of woman- and minority-owned businesses, geography, and multiple awardees.

Given that the budget for the SBIR is based on the 2.5 percent set-aside, the size of the SBIR program at DoE will change with the federal government's commitment to energy research and development. The passage of the *America COMPETES Act* in 2007, which authorizes the doubling of the budget of the DoE Office of Science, reflects growth in this commitment. As the nation's commitment to energy research increases, SBIR is expected to play a growing role in developing innovative energy technologies.

---

[1]Ref. PL102564. 10/28/1992. section 103.B modifying the section 9.F. of the Small Business Act (PL97219): "Inserting 'For the Department of Energy it shall not include amounts obligated for Atomic Energy Defense Programs solely for weapons activities or for Naval Reactor Programs'"; amends 15 U.S.C. 638 (E). The exempted programs account for 30-50 percent of the total DoE R&D budget. Exceptions include research associated with the environmental clean-up of weapons facilities and with defense nuclear nonproliferation activities.

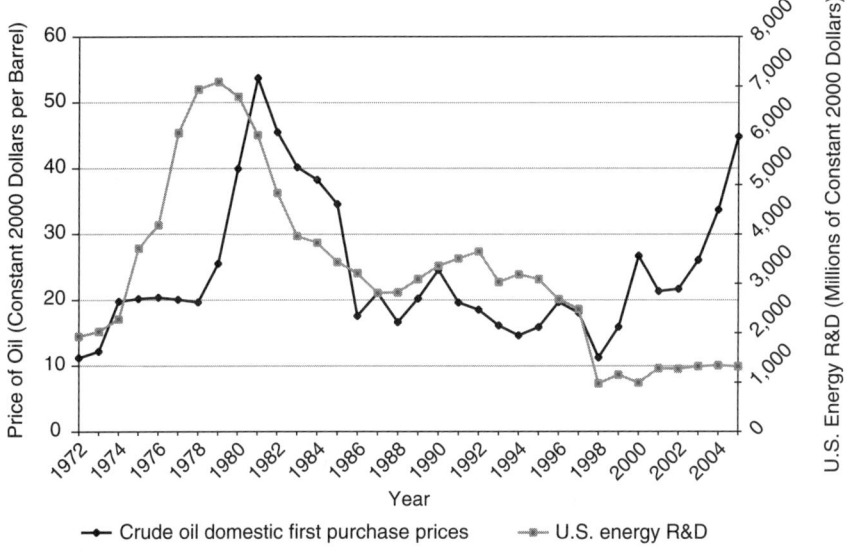

FIGURE 3-1 U.S. energy R&D and the price of oil, 1972-2005.
SOURCE: Department of Energy.

3.1 TRENDS IN ENERGY RESEARCH AND DEVELOPMENT

The federal commitment to energy research and development in the United States has undergone significant swings in the nearly 30 years since the enabling legislation creating the Department of Energy was signed into law by President Carter on October 1, 1977. Spikes in the price of oil in 1973 and 1979 drove a surge in energy R&D spending in the late 1970s and into the start of the Reagan administration. However, starting in 1981, shifting budgetary priorities favoring military technology investments and a subsequent collapse of the price of oil drove a sharp drop in commitments to energy R&D that lasted through the 1980s (See Figure 3-1). By the mid-1990s, investments in energy R&D were, in real terms, less than half their level at their 1979 peak.[2] Federal and industry R&D comprised only 0.5 percent of all sales in the industry—the lowest percentage of any industry, with the exception of primary metals.

Weak support for energy technology in the late 1990s is reflected in the data on venture capital activity. The boom in venture capital disbursements from 1995 to 2000 did result in a growth in funding by venture capital firms in energy technology companies, but at a far less-than-proportional rate. Where in 1995, disbursements to energy companies accounted for only 6.8 percent of the venture

---

[2]National Science Foundation, *Federal R&D Funding by Budget Function: Fiscal Years 2003-2005 (historical tables)*, Arlington, VA: National Science Foundation.

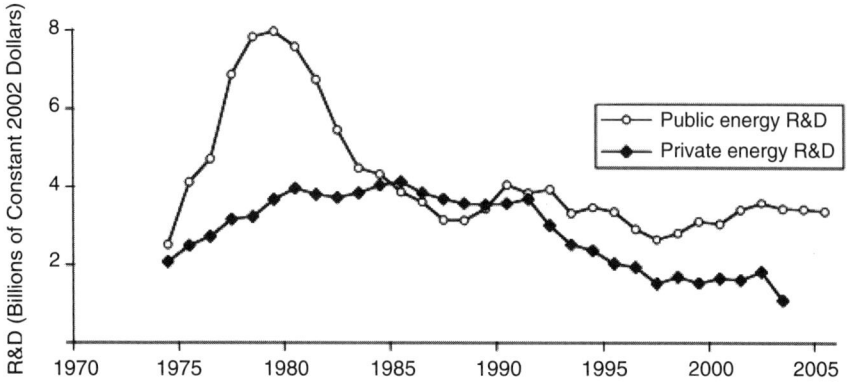

FIGURE 3-2 Public and private energy R&D.
SOURCE: Daniel Kammen, Testimony before the Committee on Appropriations, Subcommittee on Energy and Water, U.S. House of Representatives. February 28, 2007.

capital industry total, by 2000 that number had dropped to 2.4 percent. Of the $2.5 billion in venture capital invested in energy companies, the vast majority went to firms in later stages of development rather than seed stages or early stage firms.

In the past three years, energy markets have reversed course. The weak market for oil that contributed to a low level of interest in energy R&D over past decade and a half has been replaced by a very tight market in which prices have responded sharply to perceived shifts in fundamentals. Oil prices have increased dramatically. These shifts have created new opportunities for energy technology companies that are reflected in the recent venture capital numbers: In the second quarter of 2006, both the number of deals and total equity investments by venture capital firms in energy companies tripled in relation to the previous year. A growing public concern in the United States around the need to strengthen the nation's capabilities to produce energy from a variety of sources has driven a growing consensus on the need to increased investments in energy R&D. To the extent that budgetary priorities shift again, this time in favor of energy R&D, DoE SBIR program will take on an even greater role in supporting the development and commercialization of new energy technologies.

Figure 3-2 shows that public investments in energy R&D have fallen in real terms since the mid-1980s, and that private sector investments have not replaced the declining public funds.

The lack of funding is especially apparent in the limited VC interest in energy-related companies, described in Table 3-1.

TABLE 3-1 Venture Capital Disbursement to Energy-related Companies, 1995-2002

Year	Total Venture Capital Disbursements (in Millions of Dollars)	Disbursements to Energy Companies (in Millions of Dollars)	Disbursements to Energy Companies as Percent of Total
1995	7,859	535	6.8
1996	10,777	503	4.7
1997	11,644	743	6.4
1998	20,737	1,349	6.5
1999	53,415	1,666	3.1
2000	104,232	2,492	2.4
2001	40,541	1,134	2.8
2002	21,760	700	3.2
2003	19,634	811	4.1
2004	22,029	713	3.2
2005	22,640	769	3.4

SOURCE: Based on data from PriceWaterhouseCoopers, MoneyTree, <http://www.pwcmoneytree.com>.

3.2 SIZE OF INDIVIDUAL AWARDS

DoE uses the maximum dollar amounts provided by the law as its upper limits for both Phase I and Phase II awards, $100,000 and $750,000, respectively. Nearly all Phase I awards are proposed and awarded at or near the maximum. The average amount of a Phase II award is typically somewhat less than the maximum, for three reasons:

(1) applicants do not always request the maximum amount,
(2) DoE technical program offices sometimes reduce award amounts in order to fund more awards, and
(3) some of the requested amounts are reduced during negotiations conducted by DoE operations office (although DoE's SBIR office is not sent any information on the results of this process).

In FY2005, the average size of Phase II awards was $699,557.[3] Grant funding in excess of the stated maximums is rare but possible, based on the technical review of the proposals.[4]

Once the nominal size of individual Phase I and Phase II awards is fixed, the projected distribution between Phase I and Phase II awards is determined by

---
[3] U.S. Small Business Administration, Tech-Net database.
[4] This happened only twice, in 2001 and 2003, and the total number of awards would have been reduced in consequence because the allocation for SBIR funds is fixed each year. U.S. Small Business Administration, Tech-Net database.

TABLE 3-2 DoE Phase I Awards, 1992-2005

Fiscal Year	Number of Phase I Awards	Average Award Size ($)	Total Phase I Dollars
1992	196	49,765	9,753,886
1993	167	74,076	12,370,630
1994	209	74,363	15,541,961
1995	196	74,570	14,615,796
1996	167	74,795	12,490,776
1997	194	74,099	14,375,212
1998	204	74,669	15,232,440
1999	185	99,432	18,394,937
2000	292	68,152	19,900,505
2001	310	68,166	21,131,432
2002	328	68,894	22,597,310
2003	323	67,876	21,923,938
2004	257	95,325	24,498,613
2005	258	99,449	25,657,718

SOURCE: U.S. Small Business Administration, Tech-Net Database.

establishing one more parameter: the conversion ratio between Phase I and Phase II awards. DoE aims to convert 40 percent of Phase I awards to Phase II. Once this parameter is established, and the total SBIR budget is determined for a given year, a simple algebraic formula determines the number of Phase I and Phase II awards as a function of the total budget, the size of Phase I and Phase II awards, and the conversion ratio.[5]

3.2.1 Phase I Awards

The average annual number of Phase I awards was 235 from 1992 to 2005. However, there was a shift in pattern during this period. Between 1992 and 1999, DoE made an average of 190 Phase I awards; from 2000-2005 inclusive, that number jumped to 295 (Table 3-2).

Table 3-2 also shows the size of Phase I awards during this period. While total funding for Phase I awards more than doubled from 1992 to 2003, the average value of awards remained close to the SBA guidelines for the maximum award amount. The latter doubled to $100,000 by 1999, where it has remained.

---

[5] In practice, the actual success ratio for Phase II proposals exceeds the nominal conversion ratio from Phase I to Phase II because: (1) some Phase I awardees do not submit Phase II proposals, and (2) some Phase II proposals are not submitted at the upper limit of funding. Therefore, in recent years the actual success ratio for Phase II proposals has been about 50 percent.

TABLE 3-3 DoE Phase II Awards, 1992-2005

Fiscal Year	Number of Phase II Awards	Average Award Size ($)	Maximum Award Size ($)	Total Phase II Dollars
1992	66	488,010.15	500,000.00	32,208,670.00
1993	72	498,793.28	500,000.00	35,913,116.00
1994	61	596,332.13	600,000.00	36,376,260.00
1995	77	725,064.56	750,000.00	55,829,971.00
1996	70	734,996.17	750,000.00	51,449,732.00
1997	82	722,696.84	750,000.00	59,261,141.00
1998	83	741,516.16	750,000.00	61,545,841.00
1999	85	696,164.75	750,000.00	59,174,004.00
2000	91	711,715.07	750,000.00	64,766,071.00
2001	98	676,184.63	900,000.00	66,266,094.00
2002	103	694,453.72	750,000.00	71,528,733.00
2003	103	706,382.41	892,342.00	72,757,388.00
2004	115	719,960.74	750,000.00	82,795,485.00
2005	107	699,556.68	750,060.00	74,852,565.00

SOURCE: U.S. Small Business Administration, Tech-Net Database, and Department of Energy SBIR program.

DoE did not make any Phase I awards during this period above the $100,000 SBA guidelines.

3.2.2 Phase II Awards

As shown in Table 3-3, DoE has also experienced growth in the number of Phase II awards during 1992-2005. The number of Phase II awards increased gradually, by approximately 62 percent over this period. In 1992, DoE funded 66 Phase II awards, and by 2002 over 100 awards were being funded. Total DoE funding for Phase II awards more than doubled during this period, with a jump in funding in the mid-1990s. The average value of a Phase II award increased significantly in 1994 and 1995 and remained constant for the remainder of the period. Compared to Phase I, in which the average award was very close to the SBA guidelines for the maximum award amount, the average value of a Phase II award was approximately $40 thousand to $50 thousand less than the SBA guidelines for the maximum award amount, although there were two oversized awards, one in 2001 and one in 2003.

3.3 GEOGRAPHIC CONCENTRATION

The SBIR legislation, along with the SBA's Policy Directive, specifies that awards shall be based primarily on scientific and technical merit, along with

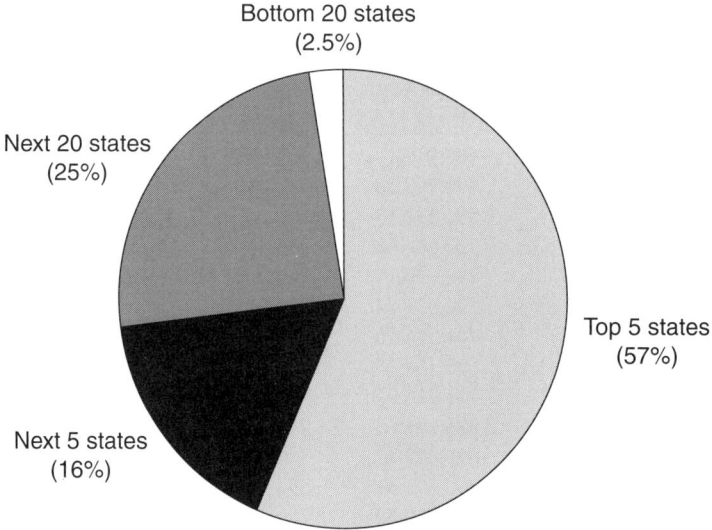

FIGURE 3-3 Distribution of DoE Phase I awards by state groupings, 1992-2005.
SOURCE: U.S. Small Business Administration, Tech-Net Database; Department of Energy SBIR Program.

considerations of a project's commercial potential. Agencies are not required to account for geographic balance in the selection of their SBIR awards.

The top five states, in terms of Phase I DoE SBIR award receipt, are California, Massachusetts, Colorado, Connecticut, Texas, and New York, and together they account for 57 percent of Phase I awards. As a point of comparison, the top five states in terms of overall R&D expenditures in 2002 accounted for 47 percent of total R&D funds. At the other end of the spectrum, the bottom 20 states captured only 2.5 percent of all Phase I awards. All but one of these states (Montana) received less than 10 Phase I awards. Three states (Alaska, Idaho, and South Dakota) and Washington, D.C., received no Phase I awards.

Applications largely mirror awards. Figure 3-4 shows the distribution of DoE Phase I applications by state groupings. The number of applications is geographically concentrated, but less so than awards. The top five states accounted for 47 percent of all DoE Phase I applications, noticeably less than the 57 percent of Phase I awards received by the top five states. On the other hand, the states at or near the bottom had a greater percentage of application than awards: the bottom 20 states accounted for 5 percent of all Phase I applications but only 2 percent of awards. Further research is required to determine why applications from the top five states have been more successful than those from the bottom 20 states.

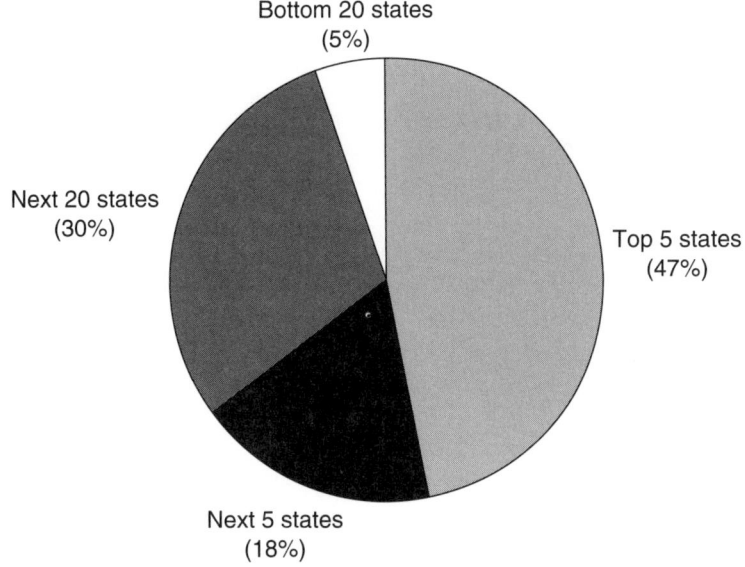

FIGURE 3-4 Distribution of Phase I applications by state, 1992-2003.
SOURCE: Department of Energy SBIR Program Web site.

Every state and territory had at least one Phase I application. Three states had at least 1,000 applications for DoE Phase I awards, with California topping the list at 3,052. Figure 3-5 shows that the distribution of DoE Phase II awards is more highly concentrated than for Phase I awards. The top five states for Phase II awards—the same as those for Phase I awards—received 60 percent of all Phase II awards. The bottom 20 states accounted for even fewer awards than for Phase I—only 1 percent of all Phase II awards. Ten states received no Phase II awards.

The distribution of DoE Phase II applications is similar to that for Phase II awards. Even within high award states, awards are clustered. The top 10 percent of Metropolitan Statistical Areas (MSAs) by Phase I awards are listed in Table 3-4.

Some of the nation's top technology regions are characterized by high number of patents or the number of venture capital recipient firms—e.g., Silicon Valley; the Boston metro area. However, others are not large cities with well-developed networks of institutions supporting technological innovation. Rather, many are college towns: Ann Arbor, MI; Boulder, CO; Tucson, AZ; Santa Barbara, CA; Madison, WI; New Haven, CT; and Albany-Schenectady-Troy, NY. Two are emerging technology regions on the periphery of the Greater Boston metropolitan area: Lawrence and Lowell, MA. A third group is home to national laboratories

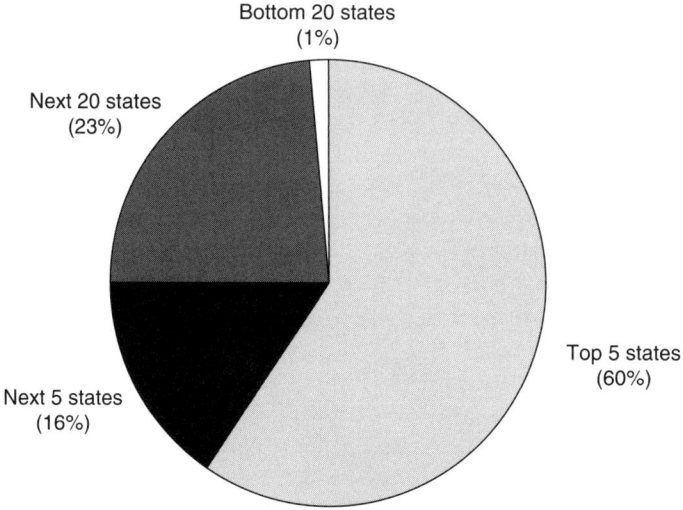

FIGURE 3-5 Phase II awards by state, 1992-2005.
SOURCE: U.S. Small Business Administration, Tech-Net Database; Department of Energy SBIR Program.

and/or major military facilities: Huntsville, AL (the Army's Redstone Arsenal housing the Army Missile Command and NASA's Marshall Spaceflight Center); Dayton, OH (Wright-Patterson Air Force Base, including a large portion of the Air Force Research Laboratory); Melbourne, FL (NASA's Kennedy Space Center); Chicago (Argonne National Laboratory and DoE's Fermi Laboratory), and Albany/Schenectady (the Air Force Rome Laboratories).

3.4 MULTIPLE-AWARD WINNERS

Multiple-award winners are in many cases exceptionally capable firms, conducting mission-relevant research, commercializing new products, and creating employment through spin-off firms.[6] Firms that repeatedly win many SBIR awards and yet generate minimal commercial products from these awards are often called "SBIR mills."

From the perspective of DoE, two core questions emerge: First, to what extent are multiple awards made to individual companies in the DoE SBIR program? Second, what outcomes are associated with these awards? We review data related to the first question in this section; the second will be addressed in the outcomes chapter of this report.

---

[6]See, for example, the case studies of Eltron Research and Diversified Technologies in Appendix D.

TABLE 3-4 Top 10 Percent of Phase I Recipient Locations, by MSA

MSA number	MSA Name	Number of DoE Phase I Awards Received by Firms in the MSA (cumulative 1992-2004)
1120	BOSTON MA-NH	1,368
7320	SAN DIEGO CA	825
4480	LOS ANGELES-LONG BEACH CA	615
8840	WASHINGTON, D.C.-MD-VA-WV	479
7400	SAN JOSE CA	466
0200	ALBUQUERQUE NM	255
1125	BOULDER-LONGMONT CO	209
8520	TUCSON AZ	198
4160	LAWRENCE MA-NH	190
7360	SAN FRANCISCO CA	183
2080	DENVER CO	149
3440	HUNTSVILLE AL	144
5120	MINNEAPOLIS-ST. PAUL MN-WI	134
0440	ANN ARBOR MI	129
2000	DAYTON-SPRINGFIELD OH	111
7480	SANTA BARBARA-SANTA MARIA-LOMPOC CA	111
1600	CHICAGO IL	93
4720	MADISON WI	79
4560	LOWELL MA-NH	78
1680	CLEVELAND-LORAIN-ELYRIA OH	77
6280	PITTSBURGH PA	75
5480	NEW HAVEN-MERIDEN CT	72
8480	TRENTON NJ	68
3280	HARTFORD CT	59
5720	NORFOLK-VIRGINIA BEACH-NEWPORT NEWS VA-NC	53
6200	PHOENIX-MESA AZ	53
6640	RALEIGH-DURHAM-CHAPEL HILL NC	53
7600	SEATTLE-BELLEVUE-EVERETT WA	48
9240	WORCESTER MA-CT	48
0160	ALBANY-SCHENECTADY-TROY NY	46
4900	MELBOURNE-TITUSVILLE-PALM BAY FL	41
5775	OAKLAND CA	40
2900	GAINESVILLE FL	36

SOURCE: U.S. Small Business Administration, Tech-Net Database; Department of Energy SBIR program.

During the period 1992-2003, 954 companies won Phase I awards from DoE. The top twenty of these companies won 546 out of 2,652 Phase I awards, or 22.3 percent of all DoE Phase I awards. Thirteen of the top twenty companies (2.6 percent of participating firms) had twenty or more Phase I awards—representing 1.7 Phase I awards per year. The average company in the top twenty received about $2.2 million in total Phase I funding. Overall, for the entire program 1983 to 2005, 1,535 companies won 4,106 Phase I awards—2.6 Phase I awards per company.)

The most prolific Phase I winner, Physical Optics Corp., received 45 Phase I awards 1992-2003, an average of 3.75 Phase I awards per year, and an average of $320,000. Physical Optics currently employs more than 100 people, many with doctoral degrees.[7]

Table 3-5 lists the top companies ranked in terms of the number of Phase II awards during 1992-2003. Again, Physical Optics is the (joint) most prolific winner, averaging 1.75 awards per year, and $1.2 million.[8] The top 20 companies accounted for approximately 30 percent of all Phase II awards 1992-2003. For Phase II, only two of the top firms received more than 20 awards, and almost half had ten or fewer awards. The average company among the top Phase II award winners received approximately $8 million in Phase II funding during the 12-year period.

The NRC Phase II Survey attempted to determine the extent to which new firms were winning awards (Figure 3-6). Twenty-eight percent of all DoE respondents reported no Phase I awards prior to the Phase I award leading to the Phase II project selected for the survey. Thirty-six percent reported five or fewer prior Phase I awards. Forty percent had no prior Phase II awards, while 36 percent reported five or fewer prior Phase II awards. Conversely, approximately 37 percent of the respondents had six or more prior Phase I awards, and about 25 percent of respondents had six or more prior Phase II awards.

For DoE respondents, the average number of prior Phase I awards was 18, and the average number of prior Phase II awards was seven. This indicates that survey responses may have come disproportionately from firms that won a larger number of SBIR awards.

3.4.1 SBIR Award Clustering to Support Technology Development

Some observers have suggested that multiple SBIR awards are used to develop more complex technologies. In the NRC Phase II Survey, companies

---

[7] According to the SBA Tech-Net database, Physical Optics acquired a total of 407 Phase I awards during this period at all SBIR agencies, amounting to $29.7 million—annual averages of 34 and $2.5 million respectively.

[8] During this period, Physical Optics won 148 Phase I awards at all agencies, amounting to $90.1 million. U.S. Small Business Administration, Tech-Net Database. This averages to 12.3 Phase II awards per year.

TABLE 3-5 Top 20 Companies Receiving DoE Phase II Awards, 1992-2003

Company	Number of Phase II Awards	Total Dollars
Physical Optics Corporation	21	14,399,812
TDA Research, Inc.	21	14,850,271
Omega-p, Inc.	18	12,650,000
Eltron Research, Inc.	18	12,099,295
MER Corporation	15	10,019,266
Radiation Monitoring Devices, Inc.	14	10,349,999
Membrane Technology And Research, Inc.	14	9,923,405
Fm Technologies, Inc.	13	9,099,404
Ceramem Corporation	13	9,200,000
Science Research Laboratory, Inc.	12	7,949,428
Calabazas Creek Research	12	7,395,865
Physical Sciences, Inc.	12	7,966,660
Tech-x Corporation	11	7,843,709
Duly Research, Inc.	11	7,699,508
Hypres, Inc.	10	6,599,415
Diversified Technologies, Inc.	10	6,436,557
American Superconductor Corporation	9	6,600,000
ADA Technologies, Inc.	9	6,299,009
Haimson Research Corporation	9	6,079,745
Bend Research, Inc.	8	4,184,884
Lynntech, Inc.	8	5,824,154
Fuelcell Energy, Inc.	8	5,249,873
Advanced Fuel Research, Inc.	8	5,136,686
Igc Advanced Superconductors, Inc.	8	4,849,929
Supercon, Inc.	8	4,599,731
Spire Corporation	8	5,494,425
Total	308	208,801,030

NOTE: More than 20 companies are listed because of tied rankings.
SOURCE: U.S. Small Business Administration, Tech-Net Database; Department of Energy SBIR program.

were asked whether they had received other SBIR awards related to the same project/technology supported by the subject Phase II award (either prior or subsequent to the subject Phase II award). Fifty percent reported having received no other Phase I awards related to that technology (excluding the Phase I award that preceded the subject Phase II), and another 42 percent indicated five or less other related Phase I awards. The average number of other related Phase I awards was two. For Phase II, 62 percent of respondents had no other related Phase II

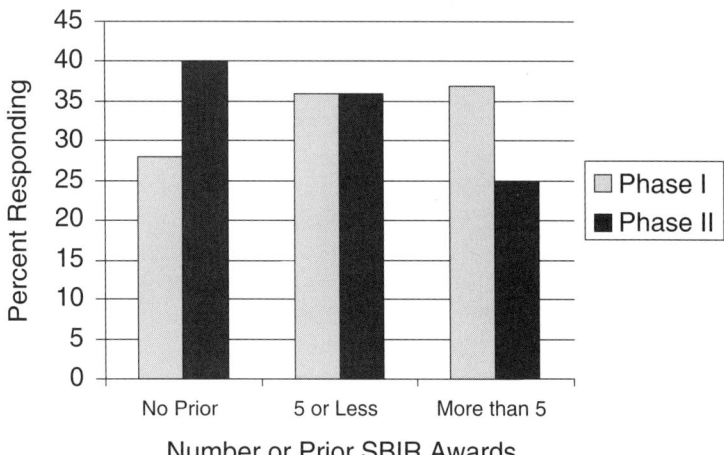

FIGURE 3-6 Number of prior SBIR awards by company.
SOURCE: NRC Phase II Survey.

awards and 37 percent reported between one and five. The average number of other related Phase II awards was one.

Case study observations in some companies (Eltron, NanoScience, Creare) also shed light on multiple awards, which can allow companies to build up a much broader and deeper technology base than would have been possible otherwise. The more complex the technology, the larger the number of complementary pieces that need to be advanced in order for the technology to be of practical use; in fact, real technological progress often occurs when a number of complementary innovations are pursued at the same time (Eltron). (However, it was also pointed out that when funding comes from multiple sources, it can be difficult for a company to attribute exact return streams to specific research projects.)

These data and cases suggest that while there are cases of clustering, it is not the predominant research structure for Phase II awardees at DoE.

3.4.2 Development Funding Prior to SBIR Award

Figure 3-7 shows the sources of investment *related to the same project/technology*, prior to the subject Phase II award. Twenty-six percent of respondents received prior SBIR funds related to technology associated with the project (excluding the Phase I award that preceded the subject Phase II).

In addition to SBIR support, other funding sources contributed to earlier R&D related to the project's technology. The largest funding came from the companies themselves: Almost 30 percent invested their own funds, including

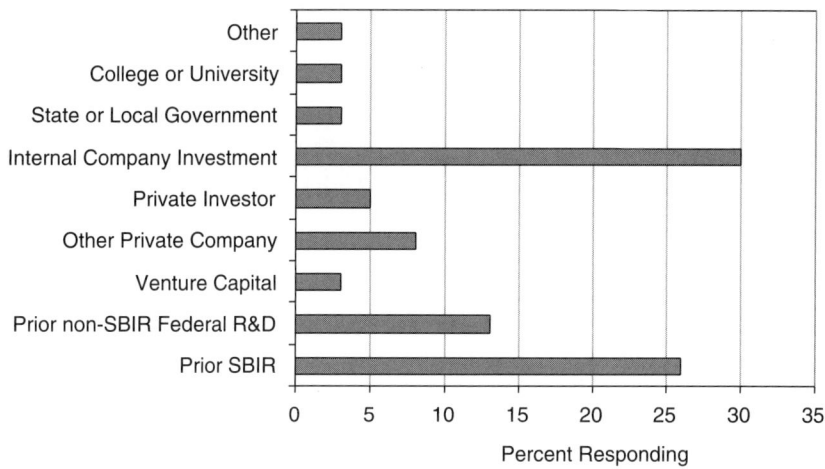

FIGURE 3-7 Sources of investment prior to Phase II award.
SOURCE: NRC Phase II Survey.

borrowed money. Approximately 13 percent of respondents had received prior non-SBIR Federal funds for related R&D. The remaining sources—such as venture capital, other companies, private investors, non-federal governments, and educational institutions—contributed in total to about 25 percent of reported cases concerning previous related R&D. However, these data do not indicate how much funding was provided by each of these sources.

4

Commercialization

4.1 CHALLENGES OF COMMERCIALIZATION

Commercialization of the technologies developed under the research supported by SBIR awards has been a central objective of the SBIR program since its inception. The program's initiation in the early 1980s in part reflected a concern that American investment in research was not adequately deployed to the nation's competitive advantage. Directing a portion of federal investment in R&D to small businesses was thus seen as a new means of meeting the mission needs of federal agencies while increasing the participation of small business and thereby the proportion of innovation that would be commercially relevant.

Congressional and executive branch interest in the commercialization of SBIR research has increased over the life of the program. A 1992 GAO study focused on commercialization in the wake of congressional expansion of the SBIR program in 1986. The 1992 reauthorization specifically "emphasize[d] the program's goal of increasing private sector commercialization of technology developed through federal research and development and noted the need to "emphasize the program's goal of increasing private sector commercialization of technology developed through Federal research and development." The 1992 reauthorization also changed the order in which the program's objectives are described, moving commercialization to the top of the list.

The term "commercialization" means "reaching the market," which some agency managers interpret as "first sale"—that is, the first sale of a product in the market place, whether to public or private sector clients. This definition, however, misses significant components of commercialization that do not result in a discrete sale. It also fails to provide any guidance on how to evaluate the scale of commercialization, an important element in assessing the degree to which SBIR programs successfully encourage commercialization.

The metrics for assessing commercialization can also be elusive, and it is important to understand that it is not possible to completely quantify all commercialization from a research project:

- The multiple steps needed after the research has been concluded mean that a single, direct line between research inputs and commercial outputs rarely exists in practice; cutting edge research is only one contribution among many leading to a successful commercial product.
- Markets themselves have major imperfections, or information asymmetries so high quality, even path breaking research, does not always result in commensurate commercial returns.
- The lags involved in the timeline between an early stage research project and a commercial outcome mean that for a significant number of the more recent SBIR projects, commercialization is still in process, and sales—often substantial sales—will be made in the future. The current "total" sales are in this case just a "snapshot half way through the race," and will require updating as the full impact of the award becomes apparent in sales.

Yet the impact of SBIR awards needs to be qualified. Research rarely results in stand-alone products. Often, the output from an SBIR project is combined with other technologies. The SBIR technology may provide a critical element in developing a winning solution, but that commercial impact—the sale of the larger combined product—is not captured in the data. In some cases, the full value of an "enabling technology" that can be used across industries is difficult to capture.

All this is to say that commercialization results must be viewed with caution, first because our ability to track them is limited (indeed it appears highly likely that our efforts at quantification of research awards may understate the true commercial impact of SBIR projects) and because an award and a successful project cannot lay claim to all subsequent commercial successes, though it may contribute to that success in a significant fashion.

These caveats notwithstanding, it is possible to deploy a variety of assessment techniques to measure commercialization outcomes. In this chapter, we review a number of metrics related to commercialization outcomes for DoE SBIR projects. These include project status, sales and licensing revenues, further investment, and employment effects.

4.2 PROJECT STATUS

Information developed from the NRC Phase II survey shows that project status varied considerably. Almost all of the DoE respondents had completed Phase II. Just over a third had discontinued Phase III activity, although nearly half

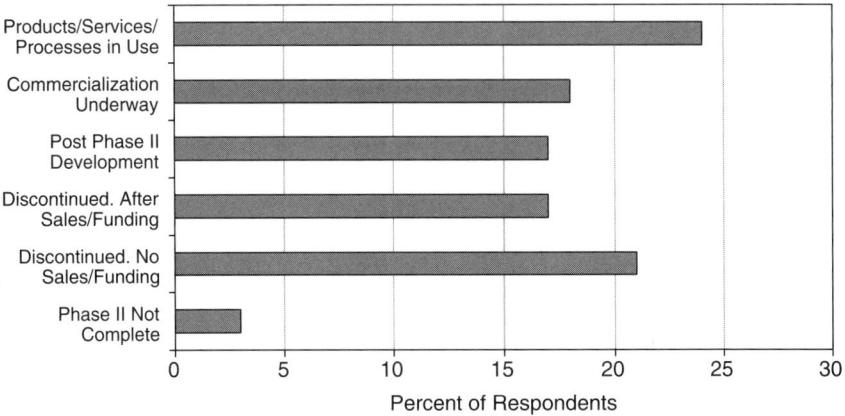

FIGURE 4-1 Status of Phase II projects.
SOURCE: NRC Phase II Survey.

of these projects had already achieved sales or other funding. Seventeen percent of responding projects were still in development; of the remaining 42 percent in the commercialization phase, just over half were already in the market place (see Figure 4-1).

4.2.1 Project Discontinuation

NRC Phase II Survey respondents cited a number of reasons for discontinuing their projects. When asked to identify the primary reason, most respondents indicated either that market demand was too small (32 percent of those discontinuing their projects), or that they had experienced technical failure or difficulties (25 percent). Not enough funding was indicated by another 12 percent.

4.3 SALES AND LICENSING

While this NRC study uses multiple indicators to assess program outcomes, and multiple metrics for most indicators, sales and licensing revenues remain an important measure of program success.

Just over half of NRC Phase II Survey respondents indicated that they had generated sales greater than $0 from their Phase II project. Averaged across all respondents (including those with no sales), the average amount of sales reported was $582,783 per project. (Of the respondents that reported positive sales, the average sales attributed to the Phase II project was approximately $1.1 million.) An additional $267,535 per project resulted from sales by licensees of the SBIR

TABLE 4-1 Reasons for Discontinuing the Project

Q: Did the reasons for discontinuing this project include any of the following?	Yes	No	Primary Reason
a. Technical failure or difficulties	41%	59%	25%
b. Market demand too small	54%	46%	32%
c. Level of technical risk too high	24%	76%	2%
d. Not enough funding	53%	47%	12%
e. Company shifted priorities	27%	73%	7%
f. Principal investigator left	10%	90%	3%
g. Project goal was achieved (e.g., prototype delivered for federal agency use)	49%	51%	5%
h. Licensed to another company	10%	90%	7%
i. Product, process, or service not competitive	31%	69%	5%
j. Inadequate sales capability	10%	90%	0%
k. Other (please specify):	7%	93%	2%

SOURCE: NRC Phase II Survey.

companies. Finally, another $48,231 per project resulted from other types of sales, including sales of the rights to technology, sales of spin-off companies, etc.

4.3.1 Skew Effects

The distribution of sales and licensing outcomes is highly skewed: Four projects accounted for 47 percent of total reported sales. These four projects each generated more than $5.5 million in sales, with one project reporting sales of $17.5 million.

It is important to emphasize that these figures are those reported by NRC Phase II Survey respondents who were asked about particular SBIR awards. The NRC review of the DoE SBIR program identified multiple other nonsurveyed SBIR-funded projects that received significant support from the DoE SBIR program and subsequently realized sales well in excess of $100 million. The Atlantia Offshore case study documents one in which cumulative product sales are over $500 million; Box 4-1 describes the case of a company (Science Research Laboratory, Inc.) whose SBIR-supported technology has driven sales of lasers ($250 million/year) used in the production of semiconductor chips.

4.3.2 Sales Expectations and Likely Future Sales

While the NRC Phase II Survey breaks new ground in collecting data to support the NRC assessment, the amount of sales made—and indeed the number of projects that generate sales—are inevitably undercounted in a snapshot survey

Box 4-1
Commercialization by Science Research Laboratory, Inc.

Science Research Laboratory, Inc. (SRL) is a high technology research and development corporation founded in 1983 by Drs. Jonah Jacob and Joseph Mangano, experts in laser technology and plasma science. The company's primary objective is to develop commercial products based upon research programs conducted for the United States government. SRL currently employs 20 individuals, more than half of whom are scientists holding doctorate degrees from some of the world's most distinguished universities. This skilled group of researchers gives the company a broad technical base, which has resulted in the development of a wide spectrum of technologies with applications in both the government and commercial sectors.

Under four Department of Energy and Department of Defense SBIR awards between 1989 and 1993, Science Research Laboratory, Inc. (SRL) of Somerville, Massachusetts developed a cluster of solid-state pulsed power technologies that made excimer lasers, for the first time, a commercially viable tool for the Deep Ultra-Violet lithography now used in writing current-generation integrated circuits onto computer chips. Specifically, these SBIR-developed technologies:

• Eliminated missing laser pulses observed with the older ("thyratron switch") technology, thereby stabilizing the laser power, improving dose control to the semiconductor wafer, and greatly improving chip yield; and
• Increased the lifespan of the laser driver by a factor of 100 and the lifetime of the laser head by a factor of 10-20, thereby reducing the annual maintenance costs of the laser fivefold.

The use of excimer lasers has enabled a reduction in the critical dimensions of the circuits from 0.35 microns to 0.25 microns with the existing laser technology, and will ultimately lead to critical dimensions of below 0.1 microns with next generation laser technology. The result has been a significant increase in the computing power of virtually every military and commercial system developed in recent years.

SRL commercialized these technologies by licensing Cymer, Inc., which went public in 1996 in part based on these technologies. As a direct result of these technologies, Cymer now produces and sells approximately $250 million annually in excimer lasers to Canon, Nikon, and ASML for use in chip production around the world and has a market share of over 80 percent.

SOURCE: Department of Defense SBIR Program.

TABLE 4-2 Recipients of Product Sales

Recipient of DoE SBIR Product	Percentage Reporting
Domestic private sector	76
Department of Defense (DoD)	2
Prime contractors for *DoD or NASA*	1
NASA	1
Agency that awarded this Phase II	0
Other federal agencies	1
State or local governments	0
Export markets	14
Other	1

SOURCE: NRC Phase II Survey.

taken at a single point in time. As noted in Box 1-1, based on successive data sets collected from NIH SBIR award recipients, it is estimated that total sales from all responding projects will likely be on the order of 50 percent greater than can be captured in a single survey.[1] This underscores the importance of follow-on research based on the now-established survey methodology.

4.3.3 Licensing

Sales resulting from licensing are not widespread among NRC Phase II Survey projects—only seven projects reported sales from licensing. Their average sales were over $6 million per project. These sales skewed the data for overall average sales by licensing per DoE Phase II Award, which averaged approximately $270,000 for all responding projects.

4.3.4 Customers

NRC Phase II Survey respondents were asked to identify the types of customers for their sales. For DoE respondents, the private sector dominated sales, with over three-fourths of sales coming from the domestic private sector and approximately 14 percent from export markets. DoE data indicates that 86 percent of sales were made to the private sector. Notably, none of the surveyed projects made any sales to DoE itself, suggesting that for SBIR at least, DoE does not operate a procurement-type program like DoD and NASA.

---

[1] Data from NIH indicate that a subsequent survey taken two years later would reveal very substantial increases in both the percentage of firms reaching the market, and in the amount of sales per project. See National Research Council, *An Assessment* of the Small Business Innovation Research Program at the National Institutes of Health, Charles W. Wessner, ed., Washington, D.C.: The National Academies Press, 2008.

TABLE 4-3 Responses Concerning Marketing Activity for Phase II Project

Marketing activity	Percent				
	Planned	Need Assistance	Underway	Completed	Not Needed
Preparation of marketing plan	4	3	15	44	34
Hiring of marketing staff	3	5	9	22	61
Publicity/advertising	9	6	26	29	30
Test marketing	8	5	20	18	50
Market research	5	5	22	34	33
Other	0	0	3	3	39

SOURCE: NRC Phase II Survey.

4.3.5 Marketing

Despite the congressional emphasis and DoE focus on commercialization, many DoE firms do not view marketing as a necessary component of the commercialization process. Thirty-four percent of DoE respondents to the NRC Phase II Survey reported that a marketing plan was unneeded, 61 percent hired no additional marketing staff, 30 percent felt publicity or advertising were unnecessary, 50 percent believed test marketing was unneeded, 33 percent had no intention of engaging in market research, and 39 percent believed other types of marketing were unnecessary (Table 4-3).

This aspect of award winner activities could bear further analysis, particularly through further work on the relationship between marketing activities and outcomes.

4.3.6 Additional Development Funding

Over 60 percent of the NRC Phase II Survey's DoE respondents received or invested additional developmental funding for the Phase II projects targeted in the survey.[2] Table 4-4 shows the distribution among the various sources of Phase III development funding. The amounts shown are the averages among all DoE respondents.

On the average, approximately $363,000 was derived from non-SBIR federal funds, nearly $225,000 from other domestic companies, $157,000 from one's own company (including borrowed money), $125,000 from foreign private investment, $37,000 from other private equity, and $13,000 from personal funds. The average amount of funding from other sources was minimal: $2,000 from state or local governments and $817 from colleges or universities. Altogether, the average total amount of additional development funding received by these DoE SBIR firms was approximately $920,000.

---

[2]NRC Phase II Survey, Question 22.

TABLE 4-4 Sources of Phase III Funding for Further Development, Following the Phase II Project

Source	Average Phase III Funding ($)
Non-SBIR federal funds	362,968
Other Private Sources	
1) Private U.S. Venture Capital	0
2) Private Foreign Investment	124,522
3) Other Private	36,908
4) Other Domestic Private Company	224,358
Other Domestic Sources	
1) State or Local Govt.	2,025
2) College or University	817
Other Not Reported	
1) Own Company	156,621
2) Personal Funds	13,216

SOURCE: NRC Phase II Survey.

The distribution of projects by funding source was somewhat different: About 40 percent reported raising internal company funds, and approximately 20 percent generated funds from other private companies and from federal non-SBIR sources (though not DoE). So while federal funding and other private companies were the largest sources of funding, they were relatively more concentrated in fewer firms.

Box 4-2
How SBIR Companies Commercialize—
Findings from the Case Studies

Companies interviewed for this study illustrate some of the many approaches to commercialization taken by DoE SBIR recipient firms. The case studies of these companies can be found in Appendix D.

One company marketed the technology developed with its lone project, subcontracted production, and achieved huge revenues (Atlantia). Another marketed and manufactured its own product (IPIX), eventually going public. Three companies are manufacturing products based on their SBIR work, while they seek larger partners in order to expand markets (NanoScience, NexTech, Thunderhead Engineering). Other companies achieve commercialization by spinning off (Creare) or licensing (Eltron) their SBIR technologies. Yet another used SBIR research to create a new market—in one instance for its core technology (Diversified Technologies, Inc.) and, in another, for an R&D contracting business (PPL).

Although some case study companies are currently dependent on SBIR as a major source of their revenues (Airak, Creare, NanoScience), all are actively engaged in commercialization, selling either products or services. Most companies use patents to protect their intellectual property.

4.4 FURTHER INVESTMENT: PHASE III AT DOE

As with other agencies' SBIR winners (and as in early stage technology development in general) DoE SBIR Phase II winners face formidable challenges in moving from the research phase to production. The period in which a company must navigate the challenge of transforming a scientific breakthrough into a market-ready prototype is often referred to as the "Valley of Death."[3] On one side of this valley stand the scientists and technologists—the innovators undertaking the research and development work. Prior to reaching the "valley," they were funded through corporate or government research funds or—more rarely—from personal sources. On the other side stand innovation managers and investors. Experts in financing and management of business enterprises, they possess development funds and expertise for turning a market-ready prototype into a validated business. But who funds the work needed to cross the Valley of Death?

The most likely possibilities are venture capitalists, large corporations, and the federal government. At other agencies, each has played an important role. However, at DoE, it appears that none has had a significant impact on Phase III activity. Instead, firms have tried to navigate the Valley of Death using internal resources or other contracts.

4.4.1 DoE SBIR and Venture Capital (VCs)

In response to the NRC Phase II Survey, no further investment was reported from U.S. venture capital sources. Various interpretations of this absence are possible.

The most obvious is that no respondents were attractive candidates for venture capital investment. This however appears to be an oversimplification. In fact, three distinct factors appear to be at work.

- First, relatively few VC firms operate in the seed capital space occupied by most SBIR awards. In the aggregate during the study period less than 5 percent of venture funding went to seed stage firms in all industries. Only at NIH, where VC activity in the biotech and other medical sectors has been substantial but appears to be a special case, have VC's played an important role in commercialization.
- Second, as noted earlier, VC interest in energy-related investments has been minimal, for many reasons. Less than 7 percent of VC investment during the study period went to energy-related companies at all stages of development. Most venture firms would rather invest in an innovation that has at least some proven

---

[3]Four case study firms raised the "Valley of Death" issue: Airak, Creare, and PPL. See Appendix D. See also the case study of Pearson Knowledge Technologies in National Research Council, *An Assessment of the Small Business Innovation Research Program at the Department of Defense*, Washington, D.C.: The National Academies Press, 2007 Prepublication.

track record, in order to minimize risk.[4] As a number of companies interviewed for this study pointed out, the venture capital market is particularly averse to funding technology projects in the early stages.[5]
• Third, firms that seek SBIR funding may be self-selected among small technology companies as ones that are less prepared to relinquish control to outside investors. Some of the case studies support this view.[6]

Recent changes in the attitude of VC firms toward energy, and apparently improved understanding among some VC firms that SBIR can help to identify especially promising technologies, may mean that in the future VC's will play a more important role at DoE. These changes also indicate that it might make sense for DoE to consider how it might help to attract VC funding and other sources of finance into the Phase III activities of its SBIR awardees. Some firms, such as NanoScience, would clearly welcome this.[7]

4.4.2 Equity Investments from Large Corporations

About 20 percent of responding projects identified additional development funding from other U.S. corporations (though no size was identified). On average, these companies provided an injection of about $675,000 per project.

Three firms interviewed for this study—Creare, Pearson, Diversified Technologies, Inc.—noted, however, that private companies are unlikely to provide the type of funding offered by SBIR due to the high risk nature of the work. Creare, for example, reported being approached by a large multinational interested in an SBIR-developed technology. The company offered to assist with marketing and distribution once the technology had been fully developed into a product; however, the company was unwilling to offer any of the development funds required to get it from a prototype to production. These firms were looking for patient investors able to take a longer-term view of R&D, yet there was a lack of optimism that this role would be filled by a large company under pressure for fast returns (PPL).

4.4.3 Other Resources

Moving from a Phase II prototype to product development and commercialization requires resources well beyond what most SBIR firms could muster

---

[4]See, for example, the case study of Airak in Appendix D. See also the case study of Pearson Knowledge Technologies in National Research Council, *An Assessment of the Small Business Innovation Research Program at the Department of Defense*, op. cit.

[5]See case studies of Eltron, IPIX, and NanoScience in Appendix D. See also the case study of Pearson Knowledge Technologies in National Research Council, *An Assessment of the Small Business Innovation Research Program at the Department of Defense*, op. cit.

[6]See the case studies of Creare and Diversified Technologies, Inc., in Appendix D.

[7]See also National Research Council, *The Advanced Technology Program: Assessing Outcomes*, Charles W. Wessner, ed., Washington, D.C.: National Academy Press, 2001.

internally.[8] And banks were viewed as a poor option because, as the firms interviewed for this study indicated, they know little about the business of small, R&D-intensive companies (NanoScience), and sought collateral beyond expectations of future knowledge and technology (Eltron).

4.4.4 Matching Funds and Cost Sharing

The limits of external funding are reflected earlier in the development process as well. Fifty-eight percent of NRC Phase II Survey respondents reported no matching funds, co-investment, or other types of cost sharing for their Phase II award. Thirty-one percent provided their own cost-sharing funds, 17 percent reported at least some cost-sharing funds from other companies, 5 percent received cost-sharing funds from a federal agency, 2 percent received some cost-sharing funds from venture capital, and less than 1 percent received cost-sharing funds from an angel or other private investment source.

4.4.5 Non-SBIR Federal Funding

With respect to government markets, it appears that the DoE SBIR program is focused almost exclusively on the private sector: Of firms responding to the NRC Phase II Survey that had some sales, 90 percent reported sales either to the U.S. private sector or to export markets. Only 4 percent reported sales to the federal government, and none to DoE.

Thus, in SBIR terms, DoE is best understood as being a nonprocurement agency: It generally does not purchase the results of SBIR-funded research. While there may be more opportunities to advertise the results of SBIR research within DoE and to the National Laboratories in particular, government markets appear to remain relatively unimportant for most DoE SBIR firms.

DoE's SBIR firms also see important limitations on federal funding. According to some companies interviewed for this study, government grants will typically help a company up to the development of a workable prototype, but not through the development of a scalable production capability and into the marketing phase (Airak, Diversified Technologies, Inc.).

4.5 EMPLOYMENT EFFECTS

Data from the NRC Phase II Survey indicates that SBIR support directly resulted in noticeable though minor employment growth among DoE respondents. On the average, approximately 1.5 employees were hired because of the SBIR award: 49 percent of respondents reported hiring one to five employees;

---

[8]National Research Council, *SBIR and the Phase III Challenge of Commercialization*, Charles W. Wessner, ed., Washington, D.C.: The National Academies Press, 2007.

46 percent reported hiring no additional employees. Five percent reported hiring six to twenty employees.

Similarly, firms on average retained 1.5 employees as a result of the Phase II award: 52 percent reported retaining one to five employees, 43 percent reported no additional employees retained, and 3 percent reported retaining 6 to 20 employees.

Respondents were also asked to compare their employment at the time of the Phase II award with current employment at the time of the survey. On average, respondents indicated growth from 32 to 54 employees. Although 30 percent of responding firms had five or fewer employees at the time of Phase II proposal submission, that number had declined to 17 percent by the time of the survey. Conversely, the percentage of respondents with more than 20 employees grew from 28 percent at the time of Phase II proposal submission to 36 percent at the time of the survey. Although this evidence does not prove that SBIR activity caused this employment growth, it does indicate that responding companies tended to expand after engaging in Phase II activity.

4.6 PHASE I COMMERCIALIZATION

While most commercialization attention has justifiably focused on Phase II projects, it is important to note that some firms have been successful commercializing right from Phase I. The NRC Phase I survey addressed firms that did not win Phase II awards.

4.6.1 Commercialization Resulting from the Phase I Projects

Naturally, most Phase I-only projects did not reach the marketplace. Many failed for reasons discussed below. But some were successful. Table 4-5 describes the companies' intentions with respect to the commercialization of their Phase I projects.

TABLE 4-5 Intentions with Respect to Commercialization

	Number of DoE Respondents	DoE Percent	Percent for all Agencies Combined
No commercialization planned	46	30	33
Software	9	6	16
Hardware	63	41	32
Process technology	45	29	20
New/improved service capability	12	8	11
Research tool	12	8	15
Drug or biologic	0	0	4
Educational materials	0	0	3

SOURCE: NRC Phase I Survey.

The distribution for DoE is similar to the distribution for all agencies combined. Percentages exceed 100 percent because respondents were asked to select all choices that applied.

Respondents were also asked whether any sales, which incorporated the technology developed during the Phase I project, had actually occurred. The results are shown in Table 4-6.

The term "other sales" in Table 4-6 refers to sales of rights to technology, sales of spin-off companies, etc. Once again, the distribution for DoE is similar to the distribution for all agencies combined. For a large percentage of projects (66 percent) no sales were expected.

The average amount of sales reported per project was $38,794, which is significantly less than the average sales for the Phase II projects reported in Section 4.3 ($582,783). Of the 18 Phase I projects for which positive sales were reported, the average sales were approximately $334,000, again significantly less than for Phase II. An additional $3,871 per project resulted from sales by licensees of the SBIR companies. Finally, another $1,878 per project resulted from other types of sales, including sales of the rights to technology, sales of spin-off companies, etc. The distribution of commercialization outcomes was even more skewed than for Phase II: 5 projects accounted for 82 percent of total reported sales.

4.6.2 Follow-on Development Funding Resulting from the Phase I Projects

Nineteen percent of respondents found additional developmental funding for the Phase II projects targeted in the survey. This figure is similar to the percentage of Phase II projects that believed the project would have gone ahead without Phase II funding, providing additional empirical support for the view that Phase II funding was a critical component in the "go" decision for about four-fifths of SBIR projects.

TABLE 4.6 Whether or Not the Phase I Project Led to Sales

	Number of DoE Respondents	DoE Percent	Percent for all Agencies Combined
No sales to date			
And not expected	103	66	65
But sales expected	23	15	15
But outcome in use	5	3	5
Sales have occurred			
Products	14	9	9
Processes	0	0	1
Services	6	4	6
Other sales	1	1	2
Licensing fees	2	1	2

SOURCE: NRC Phase I Survey.

Table 4-7 shows the distribution among the various sources of Phase III development funding. Altogether, the average total amount of additional development funding received by these DoE SBIR Phase I projects was approximately $212,000, again considerably less than the corresponding amount for the Phase II projects (Section 4.3.6).

4.6.3 Other Benefits of Phase I-only Projects

Most of the Phase I-only projects achieved no commercialization or Phase III developmental funding, but some of these produced noncommercial benefits, summarized in Table 4-8.

As shown in Table 4-8, a number of benefits ensued. Once again the distribution for DoE Phase I projects is similar to the distribution for all agencies com-

TABLE 4.7 Sources of Phase III Funding for Further Development, Following the Phase I Project

Source	Average Phase III Funding ($)
Non-SBIR federal funds	165,077
Other Private Sources	
1) Private U.S. Venture Capital	0
2) Private Foreign Investment	7,303
3) Other Private	258
4) Other Domestic Private Company	23,452
Other Domestic Sources	
1) State or Local Govt.	2,406
2) College or University	32
Other Not Reported	
1) Own Company	13,052
2) Personal Funds	742

SOURCE: NRC Phase I Survey.

TABLE 4.8 Noncommercial Benefits of the Phase I Projects.

	Number of DoE Respondents	DoE Percent	Percent for all Agencies Combined
Awarding agency obtained useful information	88	57	59
Firm improved its knowledge of the technology	138	89	83
One or more valuable employees were hired	52	34	27
Public benefits have or will accrue	45	29	17
This Phase I was essential to firm's founding or survival	17	11	13
No noncommercial benefits	9	6	8

SOURCE: NRC Phase I Survey.

bined. In some cases—firm foundation or survival—the relatively small Phase I award clearly has an impact on the firm.

The government also finds the results of the Phase I awards valuable. For 37 percent of the responding Phase I projects, the company received additional government grants or contracts related to the technology examined in Phase I. Table 4-9 shows the breakdown.

4.7 MULTIPLE-AWARD WINNERS

While commercialization is not the only metric to judge the activities either of an agency program or a recipient firm, it is one of the core objectives of the program. Data collected in the course of the NRC assessment does provide some important information.

First, data from the DoD commercialization database indicates that firms winning the largest number of awards commercialize more, on a per project basis, than do those winning the smallest number of awards (See Table 4-10.)

TABLE 4.9 Government Grants and Contracts Subsequent to the Phase I Project

	DoE Percent	Percent for all agencies combined
At least one SBIR Phase I received in this technology	26	22
At least one related Phase II—Although the subject project itself did not advance to Phase II	14	14
Subsequent federal non-SBIR contracts or grants in this technology	13	12

SOURCE: NRC Phase I Survey.

TABLE 4-10 Commercial Results from Multiple-Award Winners

Number of Phase II SBIR per Firm*	Number of Firms	Number of Projects in CCR Database	Number of Projects with Award Years Prior to 2004	Average Commercialization of Projects with Award Years Prior to 2004 ($)
≥125 projects	5	941	823	2,067,719
≥75 and <110 (no firms had between 111 and 124 projects)	5	485	411	1,117,325
≥50 and <75	17	1,067	945	4,103,125
≥25 and <50	77	2,692	2,330	1,710,140
≥15 and <25	101	1,858	1,535	1,375,061
>0 and <15	2715	8,101	6,243	1,300,886

SOURCE: Data taken from the DoD Company Commercialization (CCR) Database.

Second, the DoE case studies indicate that commercialization make take forms other than generating sales. Creare, one of the most frequent DoE winners, has been highly successful in forming spin-out companies to undertake commercialization of specific products developed with support from SBIR.

Third, data from the NRC Phase II Survey indicates that the largest commercial successes at DoE—generating more than $5 million in revenues—were all created by firms with at least 20 employees, although it is also worth noting that none came from firms with more than 300 employees.

5

Agency Mission

5.1 MANAGING A PROGRAM WITH MULTIPLE OBJECTIVES

Public Law 97-219 established SBIR with multiple objectives. The program is intended "to use small business to meet Federal research and development needs." This is usually taken to mean that the SBIR program should be aligned with the R&D needs of the sponsoring agency.

This objective, along with the other objectives mandated by Congress, must be translated into operational procedures for implementation.

The existence of multiple objectives for the SBIR program means that agencies are forced to confront the following question: In selecting technical topics, and in selecting proposals for award, how much emphasis should be placed on satisfying the mission needs of the agency versus commercialization or one of the other SBIR legislative purposes?

DoE's overarching mission is "to advance the national, economic, and energy security of the United States; to promote scientific and technological innovation in support of that mission; and to ensure the environmental cleanup of the national nuclear weapons complex."[1]

The Department has four strategic goals toward achieving this mission:

- **Defense Strategic Goal**: To protect our national security by applying advanced science and nuclear technology to the nation's defense.

---

[1]Department of Energy main Web site under "about us." Accessed at <*http://www.energy.gov/*> on May 25, 2006.

- **Energy Strategic Goal**: To protect our national and economic security by promoting a diverse supply and delivery of reliable, affordable, and environmentally sound energy.
- **Science Strategic Goal**: To protect our national and economic security by providing world-class scientific research capacity and advancing scientific knowledge.
- **Environment Strategic Goal**: To protect the environment by providing a responsible resolution to the environmental legacy of the Cold War and by providing for the permanent disposal of the nation's high-level radioactive waste.

The SBIR program is structured to support all four mission objectives of DoE in a manner that parallels the overall allocation of R&D funding to non-weapons programs.

Along with the other federal agencies that participate in SBIR, DoE attempts to strike a balance between these two purposes. However, at DoE, the link to agency mission—including the goal of performing quality science—is the program objective most clearly built into the program structure. During the October 24, 2002, conference that helped launch the NRC SBIR assessment, Milton Johnson, the Deputy Director for Operations in the Office of Science, stated that DoE Office of Science "lives and dies by the quality of the science" produced. "If we produce lousy science, soon we won't get much money for it."[2] In a May 2003 meeting, Dr. Johnson stated that SBIR was regarded within DoE like any other R&D program—that is, as a vehicle by which research programs could accomplish R&D objectives.[3] According to Johnson, the difference with SBIR was simply that the R&D work was performed by small businesses instead of national laboratories or universities.

However, Dr. Johnson also noted that the ability of small business to achieve excellent science is not the only measure of success. As a dual purpose program, the SBIR seeks not only to increase the involvement of small business in federal R&D but also to increase private sector commercialization of innovations derived from federal R&D. He concluded that the view at DoE (based on the agency's interpretation of the enabling legislation) is that commercialization will follow because it is in the best interest of the performers—the small businesses themselves.[4]

Throughout the history of the SBIR Program at DoE, the SBIR Office has promulgated a set of evaluation criteria that reflects this upper management

---

[2]Milton Johnson, "SBIR at the Department of Energy: Achievements, Opportunities, and Challenges," in National Research Council, *SBIR: Program Diversity and Assessment Challenges*, Charles W. Wessner, ed., Washington, D.C.: The National Academies Press, 2004.

[3]National Research Council symposium, "SBIR Program: Identifying Best Practice," May 28, 2005, Washington, D.C.

[4]Interview with Arlene De Blanc, SBIR Program Analyst, Department of Energy, November 3, 2003.

philosophy, as articulated by Dr. Johnson above. That is, the likelihood of commercialization is included among the criteria for evaluating grant applications, but is outweighed by other criteria, which emphasize the scientific and technical merit of the proposal.

5.2 ALIGNMENT ISSUES FOR SBIR AND THE DOE MISSION

Despite the above approach to balancing the program's dual purposes, the program has not always enjoyed unanimous support throughout all levels of management at the agency. The conceptual tension that exists between the two primary goals of the SBIR—to involve small firms in agency R&D and to fund projects with commercial potential—has resulted in a program that has earned increasing respect from program managers within DoE,[5] and yet continues to receive relatively low levels of intraagency resources for administration.

5.2.1 Research vs. Commercial Culture

Within the rigorous research culture that predominates within the Office of Science, managers of core research programs often are not in the habit of working with small businesses, nor of considering commercialization to be a priority in the evaluation of grant applications.

5.2.2 SBIR as a Tax

Furthermore (as in other agencies), those responsible for DoE R&D programs have only slowly come to view the SBIR program as something other than a "tax" that draws resources away from more valuable activities.

5.2.3 Administrative Burdens

Additional resistance also has derived from the relatively high cost and demanding nature of the SBIR program. SBIR proposals are less efficient to review, per dollar of research funded, because of the large number of proposals received, and more burdensome because of the tight deadlines.

Also, the performers of the research are different from DoE norm—requiring more outreach than for grantees at universities, for example, where the process of grant applications is better understood.

Finally, the agency's peer-review system is labor-intensive; DoE conveys information packages to at least three reviewers for every proposal, selected for their expertise in the proposal's subject matter, and retrieves them on time, or

---

[5]Particularly in the earlier stages of SBIR at DoE, this respect came grudgingly as some DoE programs felt SBIR may not fit well into programs that have broad research missions.

finds substitute reviewers. The peer-review process is now automated, however, and has presumedly resulted in greater efficiency in processing.

5.3 CHANGING PERCEPTIONS OF SBIR

There are indications that negative perceptions of the SBIR program among DoE technical staff are diminishing. Longtime SBIR staff member and former Acting SBIR Program Manager Arlene DeBlanc observed in an interview with the research team: "We make progress one [technical] program manager at a time. Once the [technical] program manager sees his needs are met, then he sees he can get value out of the program."[6] In this context, DeBlanc viewed an ultimate test of the program to be the willingness of at least some program managers to volunteer funds to SBIR in additional to the mandated set-aside.

5.3.1 Supporting Program Missions

Interviews with several Technical Program Managers and Technical Topic Managers, who are not part of the SBIR program office staff,[7] confirm that support for the SBIR program has grown within DoE: There was clear consensus among these managers that the SBIR research funded by their programs has supported program missions, provided useful outcomes, and strengthened the role of small firms in those missions.

5.3.2 Providing Research Quality

DoE has conducted two internal evaluations of the research quality of SBIR projects—one formal and one informal. The formal evaluation took place shortly after the department launched its SBIR program, at a time when considerable internal resistance to the program existed. Conducted by an independent program analysis office in the Office of Science (then the Office of Energy Research), that review was intended to compare the quality of SBIR research to other DoE-funded research, based on the assessments of DoE technical program managers. A former SBIR program manager recalls that the SBIR research was deemed by this study to be of slightly lower quality than other DoE funded research. These results were vigorously disputed by the SBIR office, which submitted a lengthy rebuttal.

The informal study was conducted by SBIR Program Manager Robert Berger in the mid-to-late 90s. Dr. Berger attempted to survey DoE technical program managers to assess quality. In this survey, the SBIR research results were com-

---

[6]Interview with Arlene De Blanc, SBIR Program Analyst, Department of Energy, November 3, 2003.
[7]Interviews conducted by the NRC in March 2005.

parable to the non-SBIR research.[8] This finding is consistent with anecdotal evidence based on interviews conducted as part of the present NRC study in March 2005 with TTMs and TPMs across several programs.

The NRC project manager survey specifically attempted to assess the perceived quality of SBIR research by asking the project managers to rate the quality of research on scale where ten is best, one is worst for each of the SBIR Phase II projects in the survey. As a baseline, the project managers also were asked to use the same rating scale to assess the average quality of research of other-than-SBIR projects conducted for the project manager's unit/office in the past two years. For DoE, the median rating was seven for both groups; the average rating was 6.80 for the SBIR and 7.24 for the non-SBIR group. (The standard deviation was 1.7 for both groups.) Further analysis showed that the result favoring non-SBIR research quality over SBIR quality is driven by outliers among the project managers who gave specific SBIR projects extremely low scores.

As another measure of research quality, the project managers were asked whether their office received more high quality research proposals than they could fund. Sixty-two percent of DoE project managers reported that there were more fundable projects than they could fund, 31 percent reported that they received about the right number of proposals, and only 8 percent reported receiving fewer fundable proposals than they could fund. The results for DoD/NASA were similar.

5.3.3 Research Impact

The project manager survey further asked whether the research conducted for the SBIR project made any difference in the way the project manager's office conducted research or pursued other research projects. The limited impact here may be related to the fact that, unlike DoD and NASA, DoE does not procure technology for its own use; therefore, the impact of the research may be harder to identify.

Of the 42 project managers who reported that SBIR-funded research had affected the manner in which their office had conducted research or pursued other research projects, about half encouraged the project performer to seek additional SBIR awards; more than half tried to follow up on project ideas in other research conducted or sponsored by the project manager's office; some of these further efforts led to a blind alley.[9]

---

[8] These internal findings are supported by an external review as well. In January 1989, the GAO reported that the quality of DoE's SBIR research was comparable to non-SBIR research. U.S. General Accounting Accounting Office, *Federal Research: Assessment of Small Business Innovation Research Programs*, GAO/RCED-89-39, Washington, D.C.: U.S. Government Printing Office.

[9] Note: the total projects in the three categories exceed 42 because the project managers were instructed to assign as many categories.

5.3.4 Comparative Research Value

Finally, in the survey project managers were asked to compare the value of SBIR research, per dollar spent, to that of non-SBIR research. DoE project managers rated 17 percent of the SBIR projects as providing more benefits than other agency research projects, 52 percent providing the same benefits, and 31 percent providing fewer benefits. The positive score was similar to that at DoD and NASA projects.

The project managers also were asked about two types of outcomes: (1) commercialization; and (2) other noncommercial, perhaps intrinsic, uses. At DoE, the project managers reported that 30 percent of projects were commercialized and that 54 percent had noncommercial/intrinsic use. By comparison, the DoD/NASA project monitors reported 35 percent and 68 percent respectively. Overall, project managers indicated that nearly 60 percent of DoE SBIR projects showed a commercial or intrinsic use or both.

At DoE, only a few projects were identified as receiving Phase III funding from the agency (13 percent). None of the yes responses further indicated that the product of the SBIR award had been directly procured by the agency.

5.3.5 Project Ownership

Participation in topic generation is one indication of whether a project manager would claim "ownership" of projects resulting from that topic. For about 70 percent of DoE projects, the project manager indicated involvement in the generation of the topic that led to SBIR award; 58 percent reported involvement with the technology before Phase I began, with another 30 percent reporting involvement after Phase I but before Phase II began. Overall, we can define an "ownership group" as those project managers who had a potential stake in the project, as demonstrated either by involvement in topic generation or involvement with the project before the Phase I proposal. For DoE, more than three-fourths of the survey responses had project managers that could be defined as being in the ownership group.

All but two DoE project managers reported a technical role in the project. In addition, about one-fifth reported a financial role. It is likely that the financial role was related to such decisions as (1) determining whether the costs of the project were appropriate, (2) deciding whether to fund all or part of the Phase II proposal, and (3) advising the contracting officer on whether sufficient work had been accomplished to justify payments. Only 7 percent reported having a role with respect to commercialization assistance.

Finally the project managers were asked if they or others played a role in the project's ability to obtain Phase III funding. Most DoE project managers (81 percent) did not know enough about this subject to answer the question.

5.4 CAPITALIZING ON PROGRAM FLEXIBILITY

5.4.1 Balancing Commercialization and Mission Orientation

The scoring system for SBIR at DoE is intentionally set up to provide flexibility to DoE technical programs with different priorities. A program that emphasizes basic research—for example, Basic Energy Sciences or High Energy and Nuclear Physics—can select proposals for funding that do not receive a strong endorsement on the impact criterion (provided that the criterion is not rated as "having reservations"). Similarly, a program that emphasizes the application of technology—for example, Fossil Energy or Energy Efficiency and Renewable Energy—can select proposals that do not receive a strong endorsement with respect to scientific quality.[10] For both of the examples, however, the proposal would have to receive a strong endorsement with respect to the other two criteria, in order to be considered for funding. While the quality of all proposals must be high to be recommended for award, the technical programs now have the freedom to select any of their candidates for funding for award, and are not forced to select proposals in order of score; i.e., program priorities can now be used as an important factor in award selection.

Flexibility also allows the technical programs to develop their own strategies for utilizing the capabilities of the small business performers. For example, interviews with staff indicate that the High Energy and Nuclear Physics (HENP) program employs a strategy of using SBIR to develop instrumentation to support the large scale facilities it manages. This strategy exists because they believe that this is the one place where small businesses can contribute—most of HENP research is big science, usually collaborations among large numbers of physicists worldwide, and therefore not amenable to small businesses. The Office of Biological and Environmental Research (BER) also employs SBIR for instrumentation development, frequently soliciting proposals in this area and adopting new measurement tools and methods developed by SBIR participants.

DoE technical staff capitalizes on program flexibility in other ways that seek to make the most of complementarities between research programs. Basic Energy Sciences (within the Office of Science), for example, recruits Energy Efficiency and Renewable Energy (EE) personnel to write topics of mutual interest (i.e., topics that tend to have a basic research slant, yet are applicable to EE's more applied interests). BES puts up their SBIR money; EE puts up the labor to write the topics and manage the proposals. Both programs benefit: BES partners with the applied programs (a specific goal of the program) and EE is able to fund additional SBIR projects.

---

[10]In contrast, under the scoring system originally in place for the SBIR program, proposals that received mid-range scores on any criterion were highly unlikely to be funded.

5.4.2 Internal Reallocation of Topics Among Programs

The mandate from Congress is for 2.5 percent of the federal R&D to be allocated to small businesses as part of the SBIR program. Each of the agencies has the freedom to implement this program as appropriate to their mission and structure. DoE has since the mid to late 1990s allocated the SBIR award funds on a prorated basis according to the funding of each of its 12 internal program areas.

However, some program areas have generated more applications than others. The chance of winning a Phase I award thus varied greatly between programs, as can be seen in the 2005 Award Rates (see Table 5-1). For example, a Phase I SBIR application in the Nuclear Physics program area had a 53 percent probability of success, while that of an application in Energy Efficiency and Renewable Energy was only 8 percent. On average, 19 percent of DoE SBIR applications in 2005 were successful.

These differences in award rates have already started to drive internal reallocation of funds. Some Basic Energy Sciences topics have been reallocated to Energy Efficiency and Renewable Energy. This internal reallocation occurs case by case, and is negotiated between programs. Some programs also share topics that jointly apply to both programs. However, it seems possible that this might be an area worth further examination by DoE as it seems to ensure that the best quality proposals are funded.

TABLE 5-1 2005 SBIR/STTR Award Rates

Program Area	Number of Topics	Number of Applications	Number of Phase I Awards	Award Rate (%)
Fossil energy	7	247	29	12
Advanced scientific computing research	3	47	9	19
Basic energy sciences	9	247	56	23
Biological and environmental research	6	182	47	26
Environmental management	0	0	0	0
Nuclear physics	4	47	25	53
High energy physics	5	111	46	41
Fusion energy sciences	3	80	18	23
Nuclear energy	1	11	3	27
Energy efficiency and renewable energy	6	470	38	8
Nonproliferation and national security	3	61	11	18
Electric transmission and distribution	2	48	8	17
TOTALS	49	1,551	290	19

SOURCE: Department of Energy SBIR program Web site.

6

Woman- and Minority-Owned Businesses

Support for women and minorities in federal R&D is a core objective of the SBIR program.

6.1 WOMAN-OWNED BUSINESSES

Figure 6-1 shows the share of all Phase I awards made to woman-owned businesses 1992-2005. Woman-owned businesses accounted for 7.7 percent of all DoE Phase I awards during this period. During 1992-2003, woman-owned businesses, accounted for 10.5 percent of Phase I applications.

The average size of Phase I awards to woman-owned businesses was similar to that for all recipients combined, near the nominal maximum funding limit. This indicates that woman-owned businesses did not receive differential treatment with respect to the average amount of award.

Figure 6-2 shows the percentage of Phase II awards received by woman-owned businesses in 1992-2005.

The percentage of Phase II awards received annually by woman-owned businesses remained fairly constant during this period, with minor variation from year to year. Woman-owned businesses accounted for 7.8 percent of Phase II applications in 1995-2003 and 6.4 percent of actual Phase II awards 1992-2005. Although, the average value of a Phase II award for woman-owned businesses declined from 1996 to 2000, it has since rebounded upward. The average value of awards to woman-owned businesses was comparable to those for all awardees; indeed, for approximately six years, the average value of a Phase II award was higher for woman-owned businesses than for all awardees combined.

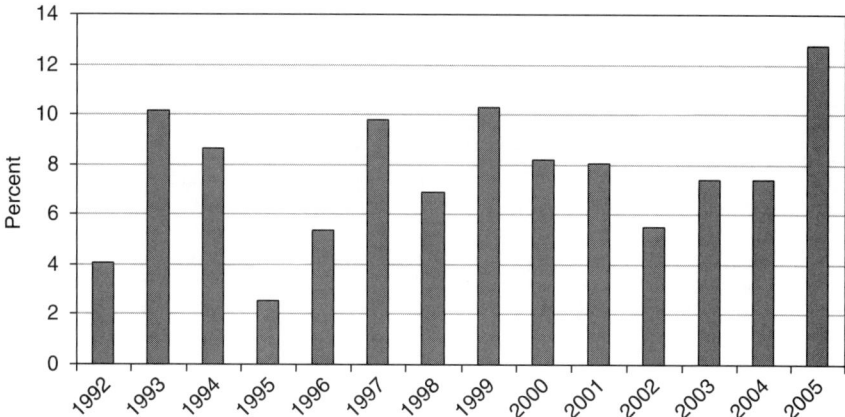

FIGURE 6-1 Phase I awards to woman-owned businesses as a percentage of all DoE Phase I awards.
SOURCE: U.S. Small Business Administration, Tech-Net Database.

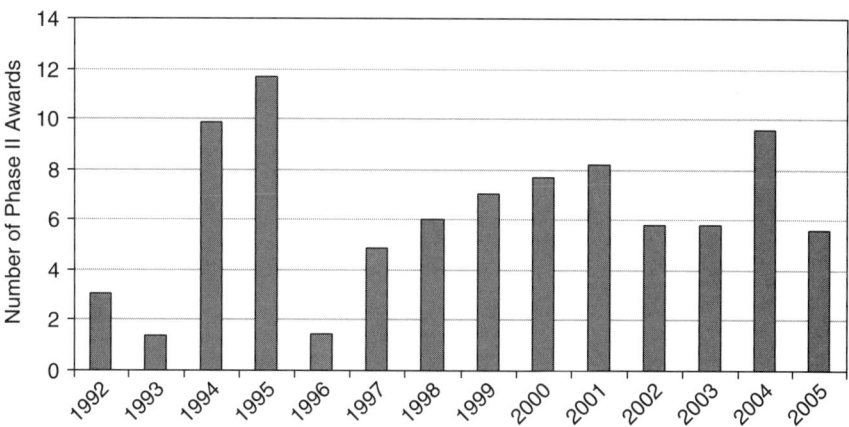

FIGURE 6-2 Phase II awards received by woman-owned businesses in 1992-2005 as a percentage of all DoE Phase II awards.
SOURCE: U.S. Small Business Administration, Tech-Net Database.

6.2 MINORITY-OWNED BUSINESSES

Figure 6-3 shows that minority-owned businesses on the whole won a significantly larger share of DoE Phase I awards than woman-owned businesses. From 1992-2005, they won an average of 13.2 percent of awards, compared to 7.7 percent for woman-owned businesses. The average value of Phase I awards

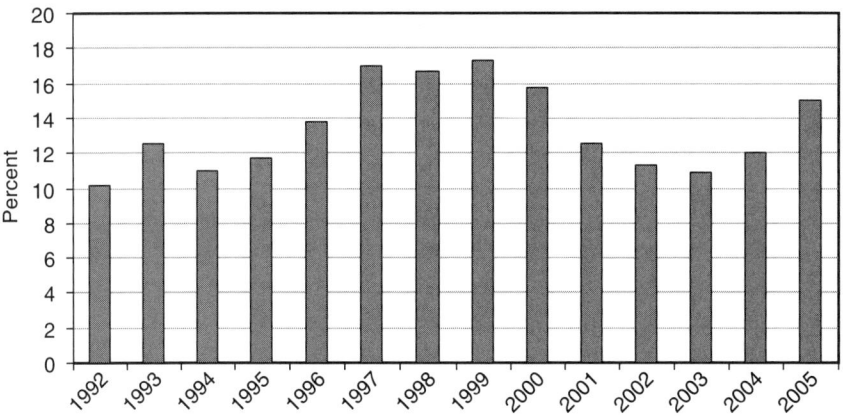

FIGURE 6-3 Phase I awards to minority-owned businesses as a percentage of all DoE Phase I awards.
SOURCE: U.S. Small Business Administration, Tech-Net Database.

received by minority-owned businesses was similar to that of other firms. In fact, in eight of the years during 1992-2003, the average Phase I award for minority-owned businesses was slightly greater than for all firms.

Minority-owned businesses submitted 16.5 percent of all Phase I applications in 1992-2003 and received 13.2 percent of all Phase I awards. Again, as with woman-owned firms, this indicates that these firms were less likely than average to generate successful applications. DoE does not appear to have analyzed these data further in order to determine why this might be the case.

The number of Phase I awards to minority-owned businesses jumped approximately 30 percent in 1997 but steadily declined from 1999-2002. It is not clear why this sharp rise and the subsequent decline occurred. Additional research might help clarify this evolution.

Figure 6-4 shows the number of Phase II awards received by minority-owned businesses in 1992-2005.

Minority-owned businesses accounted for 13.7 percent of Phase II applications in 1992-2003 and 12.1 percent of actual Phase II awards. The average value for these awards varied, but over the entire period, the average value of the awards to minority-owned was higher than for all awardees combined.

6.3 SUCCESS RATES FOR THE DIFFERENT GROUPS

For both woman-owned businesses and minority-owned businesses, and for both Phase I and Phase II awards, the success rates (percentage of grant applica-

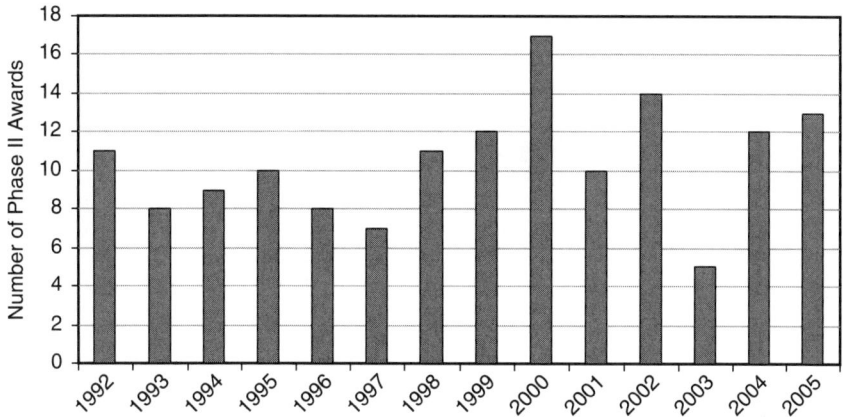

FIGURE 6-4 Number of Phase II awards received by minority-owned businesses in 1992-2005.
SOURCE: U.S. Small Business Administration, Tech-Net Database.

tions that receive awards) were lower than for all other applicants. This result is summarized in Figures 6-5 and 6-6 for Phase I and Phase II, respectively.

Figure 6-5 shows the differential rates of success by ownership status for Phase I. Woman-owned businesses have had a lower rate of success compared to all other groups—by approximately 3-10 percentage points—in every year except one. For minority-owned companies, the success rate is better than for woman-owned companies, but still lags behind the "other" category (neither woman-owned nor minority-owned). During 2002-2003 the success rate of minority-owned businesses was considerably lower than that for woman-owned and all other businesses.

Success rates for Phase II awards are similar to those for Phase I (Figure 6-6). During the period 1995-2003, woman- and minority-owned businesses lagged behind all other business in their relative success at winning Phase II awards. For the whole period, the average success rate for woman-owned businesses was 0.44 (±0.12 std. dev.) percent compared 0.52 (± 0.6 std. dev.) percent for all other businesses. In 1996, the success rate for woman-owned businesses dipped below 30 percent; however, this rate has improved—in the early 2000s, it converged towards the rate for all other businesses. The same cannot be said for minority-owned businesses. Although minority-owned businesses had a higher average success rate, 0.43 (0.10 std. dev.), than woman-owned businesses during this period, in most of the years, the success rate for minority-owned businesses was below the "all others" group. In particular, the success rate for minority-owned companies in 2003 fell sharply—to approximately 27 percent—when the rate for all other businesses was over 50 percent.

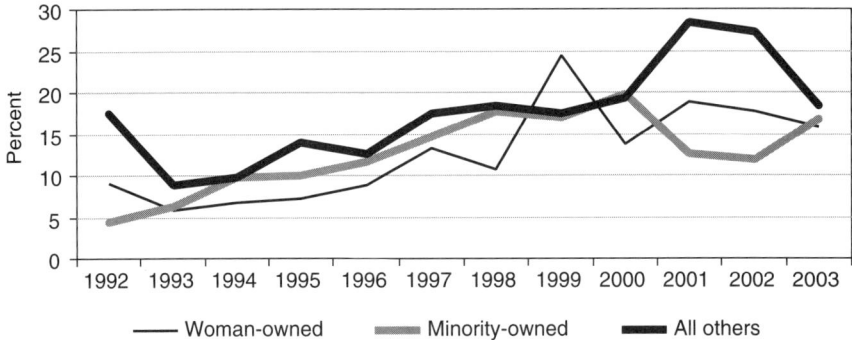

FIGURE 6-5 DoE Phase I success rates by ownership status (number of awards/number of applications).
SOURCE: U.S. Small Business Administration, Tech-Net Database; Department of Energy SBIR program.

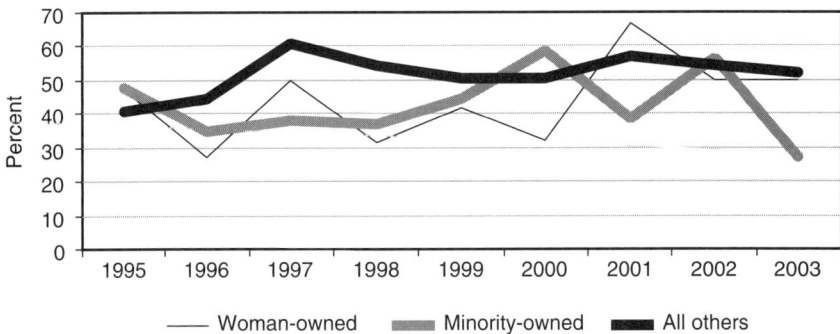

FIGURE 6-6 DoE Phase II success rates by ownership status (number of awards/number of applications).
SOURCE: U.S. Small Business Administration, Tech-Net Database; Department of Energy SBIR program.

7

Knowledge Effects

One of the congressionally mandated objectives for the SBIR program is to increase scientific and technical knowledge. The program does this through stimulating new research, building partnerships, and creating networks for the exchange of information. Patents, copyrights, trademarks, and scientific publications reflect, to varying extents, the growth in scientific and technical knowledge developed as a result of SBIR awards.

7.1 PUBLICATIONS AND INTELLECTUAL PROPERTY

Publications and formal intellectual property protection (i.e., as opposed to trade secrets) are the most easily measured of project outcomes related to technological innovation. NRC Phase II Survey respondents indicated that 123 patents related to the award had been applied for and 91 granted. Respondents had also applied for 21 copyrights and 31 trademarks, and were granted 20 and 27, respectively. Forty-three percent of DoE projects generated at least one patent application; the largest number of patents granted for an individual project was 13. Two firms were the top patent winners, with 13 each.

Peer-reviewed publications are another important indicator that the research funded by SBIR is both high quality and is reaching the scientific and technical community. Respondents indicated that 218 scholarly works were submitted for publication, or approximately 1.5 articles per Phase II project. These submissions generated 200 publications, or 1.4 published articles per Phase II project, with almost exactly half of all projects generating at least one submission, and the largest number of an individual firm being 12. This evidence suggests that

DoE Phase II research leads to outcomes of a sufficiently high standard to merit scholarly publication and granting of intellectual property protection.

Finally, the respondents to the NRC Phase I Survey were asked to provide the number of patents, copyrights, trademarks, and/or scientific publications for the technology developed as a result of the subject Phase I project (Table 7-1).

As seen from Table 7-1, the percentages for DoE are similar to those for all agencies combined.

7.2 STIMULATING NEW RESEARCH

Responses among DoE participants to the NRC Phase II Survey strongly suggest that most Phase II projects would not be implemented without SBIR funding (see Figure 7-1).

TABLE 7-1 Patents, Copyrights, Trademarks, and Scientific Publications Derived from the Phase I Project

	DoE Applications as Percent of Responses	DoE Successful Applications as Percent of Responses	Percentages for All Agencies Combined
Patents	25	19	23 / 18
Copyrights	1	1	4 / 3
Trademarks	3	3	4 / 3
Scientific Publications	45	38	38 / 35

SOURCE: NRC Phase I Survey.

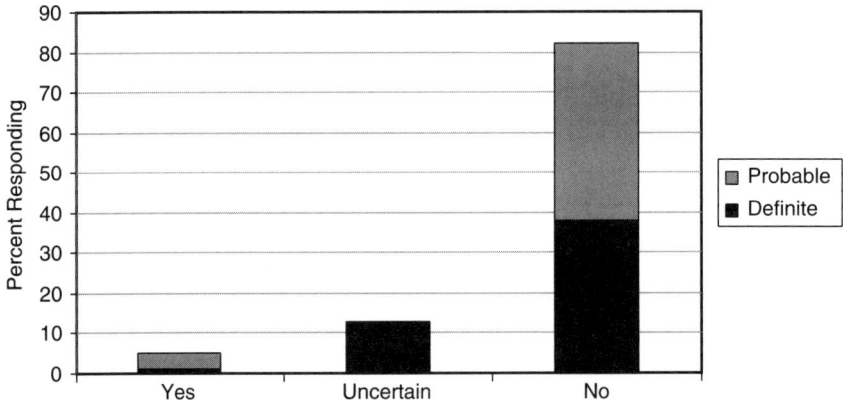

FIGURE 7-1 Would project have been undertaken without an SBIR award?
SOURCE: NRC Phase II Survey.

Thirty-eight percent of respondents reported that the project funded by the Phase II award definitely would not have been undertaken in the absence of the SBIR funding, and another 44 percent indicated that it probably would not have been undertaken. This matches data from the NRC Phase I Survey noted earlier—that about four-fifths of projects would not have been undertaken without the Phase II funding. Another 13 percent of respondents were uncertain. Only 4 percent of respondents likely would have undertaken the project without the Phase II award, and only one firm indicated that it definitely would have pursued its project.

These data are further supported by case study analysis. Without exception, all of the case study companies indicated that SBIR was vital to the development of their technology. Most said that the technology would not have been created if there had been no SBIR program.

7.3 BUILDING PARTNERSHIPS AND ENHANCING NETWORKS

The SBIR program also facilitates technological innovation through the creation of new research and commercialization partnerships and by strengthening networks of small businesses and innovators. Some of the firms interviewed for this study—for example, Creare—reported that pre-proposal networking involving small and large companies increases the likelihood of commercial sales. One company (NanoScience) found that the application process itself could become a networking vehicle for the purpose of identifying downstream partners; as the company gathered information for completing its proposal, it contacted a wide variety of industry representatives.

The active engagement of an agency program manager can also support this type of networking; in one instance, a Navy technical topic manager introduced an awardee to over 300 people and helped set up 100 presentations (Creare).

Figure 7-2 shows the type and status of relationships with other organizations, as reported through the NRC Phase II Survey. Three quarters of respondents reported either a formalized relationship with other companies and investors, or an attempt to establish such relationships, as a result of the technology developed during the subject Phase II project.

Fifteen percent had finalized plans for licensing agreements with respect U.S. companies and investors, and another 15 percent were in negotiations. Other finalized agreements also were reported: market and distribution (9 percent), customer alliances (8 percent), R&D agreements (8 percent), manufacturing (6 percent), and sale of technology rights (5 percent). In these categories, similar numbers were reported to be in ongoing negotiations, as well.

NRC Phase II Survey responses further indicate that relationships are forged not only with domestic partners but also with foreign companies or investors. Approximately 14 percent of respondents were involved in licensing agreements with foreign partners (5 percent finalized, another 9 percent in negotiations),

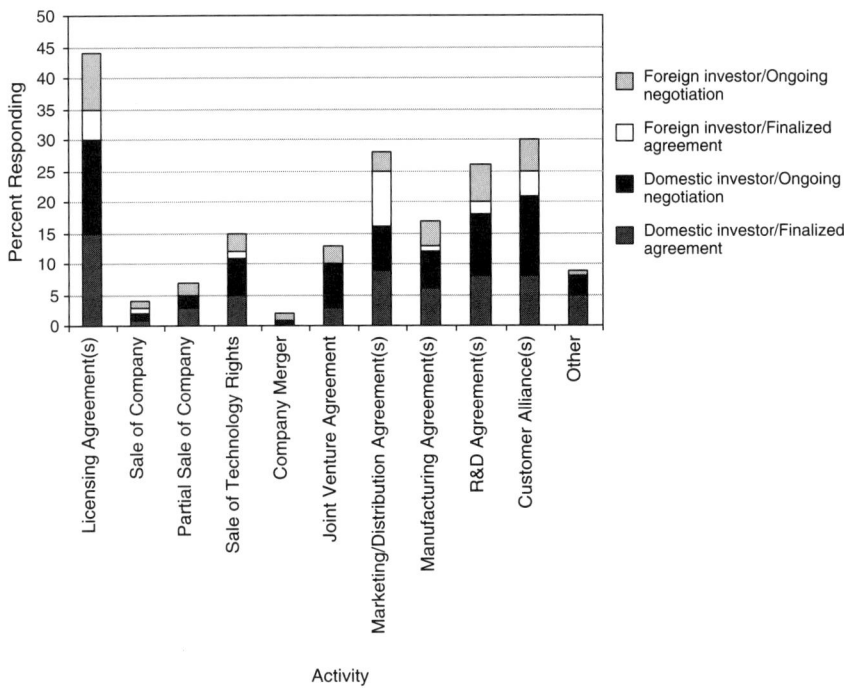

FIGURE 7-2 Company activities with other companies and investors.
SOURCE: NRC Phase II Survey.

12 percent were involved in marketing or distribution agreements, 9 percent in customer alliances, and 8 percent in R&D agreements.

7.4 SBIR AND THE UNIVERSITIES

One of the key features of SBIR programs is that they support alignments between the private sector, the agencies, and universities. This is true at DoE, as collaboration with academic research groups is common among DoE SBIR companies.

About a quarter of NRC Phase II Survey respondents reported the involvement of faculty or adjunct faculty in their project, 20 percent the involvement of graduate students, 16 percent the use of university or college facilities, and 18 percent use of a university or college as a subcontractor on the project. Five percent of respondents said that the Phase II project was originally developed at that institution by one of the recipients. Three percent reported the PI was an adjunct faculty member.

However, only one respondent reported that the technology used in the project was licensed from the university.

More generally, the NRC Firm Survey, which did not differentiate between agencies, indicated that two-thirds of responding firms have one or more firm founders who were university staff, and that 36 percent of founders were most recently employed at a college or university.

8

Program Management

8.1 SBIR IN THE DEPARTMENT OF ENERGY[1]

The Department of Energy (DoE) SBIR program is located administratively within its Office of Science (SC). With a budget of $3.5 billion, SC is the department's largest science-funding component. The programmatic emphasis within the Office of Science is on basic energy sciences, biological and environmental research, fusion energy sciences, high energy and nuclear physics, and computational science. Within the federal government, the Office of Science is the largest federal sponsor of materials and chemical sciences research.

In addition to SC, six other technical programs at DoE receive services from the SBIR office: Fossil Energy, Energy Efficiency and Renewable Energy, Electricity Delivery and Energy Reliability, Nuclear Energy, Environmental Management, and Nuclear Nonproliferation. SC is the largest DoE program participating in SBIR (receiving 64 percent of SBIR funds in 2002). As noted above, most nuclear weapons programs are excluded from SBIR.

At the outset, there were competing interests as to which DoE office should house the management of the SBIR program. Ultimately, Alvin Trivelpiece, then the Director of the Office of Energy Research, OER (the forerunner of the Office of Science, SC), persuaded the Secretary of Energy to assign SBIR responsibilities to OER. Two major factors contributed to this decision: (1) approximately two-thirds of DoE's SBIR funds would come from OER (because most of DoE's

---

[1] Information in this section is drawn from discussions with current and former DoE SBIR staff and other DoE personnel involved in DoE SBIR, and from the presentation by Milton Johnson (DoE Office of Science Deputy Director for Operations) at the October 24, 2002, National Research Council workshop, *The Small Business Innovation Research Program: Measuring Outcomes*, Washington, D.C.

Defense Programs area was exempted by law from SBIR), and (2) OER's basic research culture was thought to be more aligned with what was thought to be the spirit of SBIR research—namely, high risk/high reward.

Trivelpiece designated Ryszard Gajewski to serve as the SBIR Program Manager in addition to his responsibilities as director of OER's Advanced Energy Projects (AEP) Division. Gajewski formed an SBIR Advisory Panel to develop processes for implementing the legislation and to oversee the conduct of the program at DoE.[2] The panel provided advice to its Chairman, the SBIR Program Manager, who was able to exert considerable influence. Many of its early decisions became policies that are still in effect today. Box 8-1 summarizes major events in the evolution of the program.

Today, the DoE SBIR Office is led by the SBIR Program Manager, appointed by the Director of the Office of Science. The management of DoE SBIR program is centralized along some dimensions, decentralized along others.

- **Centralization**. Following guidance provided by the SBIR program manager, all participating technical programs adhere to a common schedule involving one competition per year, observe common procedures for the receipt and evaluation of grant applications, and follow the same guidelines with regard to scoring. The processing of proposals (officially, proposals are referred to as "grant applications" at DoE), including management of the review and selection processes, is administered by the SBIR office within the Office of Science.
- **Decentralization**. The technical programs are responsible for generating technical topics for the annual solicitations, selecting reviewers for each proposal, scoring the proposals, and recommending proposals for funding. These procedures, which are overseen by the SBIR Program Manager, are performed by approximately 70 Technical Topic Managers (TTMs) and Technical Project Monitors (TPMs) located within the technical programs that participate in SBIR.

Each program at DoE that participates in SBIR proposes topics to be included in the annual solicitation. The quantity of funding allocated to technical challenges identified by that program directly determines the number of topics included.

By design, the share of topics for each program is about equal to its share of the overall DoE extramural research budget. The SBIR program manager assigns each program a proportional allotment of technical topics for the annual SBIR solicitation, as well as an allotment of both Phase I and Phase II awards,

---

[2]The SBIR Advisory Panel was composed of one representative from each technical program that would contribute to the SBIR set-aside, plus representatives from general counsel, procurement, and budget. Typical representatives were working-level scientific program managers.

> **Box 8-1**
> **Major Events in the Evolution of the DoE SBIR Program**
>
> ***1983—DoE becomes the first agency to issue an SBIR solicitation with explicitly specified research topic areas.*** The first solicitation had 25 topics and attracted 1,734 proposals. However, the distribution was far from uniform. One topic received 342 proposals, others less than 10. DoE reacted by increasing its emphasis on topic editing, attempting to narrow some topics and broaden others.
>
> ***1986—Samuel Barish appointed SBIR Program Manager.*** From the beginning Dr. Gajewski was performing two functions (he was also AEP director) and delegated most day-to-day operations to Dr. Barish. This appointment largely formalized the status quo.
>
> ***1988—Commercialization Assistance Program (CAP) begins.*** DoE was the first agency to initiate a program to assist awardees with the commercialization of SBIR technology. Five applicants responded to a solicitation, and the competition was won by Dawnbreaker, a small company in Rochester, NY. Dawnbreaker has provided these services to DoE SBIR awardees over the next 17 years, winning several more DoE competitions. Other agencies have since adopted such programs for their own SBIR awardees. DoE believes that DoE CAP helped drive the change in the law (P.L. 102-564) that allowed agencies to use funds from the SBIR set aside for such purposes.
>
> ***1993—STTR program implemented with first solicitation.*** With far fewer funds in STTR, DoE limited the number of technical topics, implementing a topic rotation among its technical programs.
>
> ***1994—Criteria for commercial potential added to evaluation of Phase II proposals.*** In order to increase the emphasis on commercialization, P.L. 102-564 required agencies to consider four additional criteria in the evaluation of Phase II proposals. DoE formally added these criteria to its scoring system.
>
> ***1995—SBIR Process Improvement Team recommends some significant changes.*** An SBIR Process Improvement Team (PIT) was established in response to two primary driving forces: (1) the by-now entrenched tension between the SBIR office and DoE's technical programs and (2) the Office of Science initiative on Total Quality Management, pushed by its Director, Martha Krebs. The PIT, composed of five TTMs/TPMs and chaired by Robert Berger, SBIR Program Officer, was charged with developing strategies for streamlining the program and reducing tensions.

based on that program's "contribution" to the SBIR budget.[3] Once the allocation of technical topics is made by the SBIR staff, the technical program managers are free to assign topic responsibility among their own staff. Reasons for such

---

[3]The allocation of SBIR funds in proportion to R&D "contributions" is an arrangement devised in recent years that has successfully countered previous "gaming" by technical topic managers (TTMs) and technical project monitor (TPMs) in the scoring of SBIR proposals.

> **1996—Robert Berger appointed SBIR Program Manager.** Dr. Berger implemented many of the changes recommended by the PIT on which he was a member. First, measures were taken to reduce the workload: the Phase II early decision program was terminated, topics were further narrowed to reduce the number of proposals, some rules were relaxed, and paperwork was reduced. A new scoring system was instituted to provide primary authority for award selection to the technical programs; although the SBIR Program Manager still maintained oversight on scoring, the Application Score Verification Panel (ASVP) was dropped and a formal process was implemented to allow the technical programs to appeal SBIR office decisions to a neutral third party. Finally, a policy was implemented to attempt to provide all technical programs with a fair return on their SBIR investment.
>
> **1999—DoE issues a single solicitation for both SBIR and STTR.** To reduce the workload further, a single solicitation was issued for both SBIR and STTR and the two programs were put on the same evaluation schedule. To increase their chances of winning an award, small businesses that partnered with a research institution were allowed to apply to both programs with a single proposal by checking a box on the cover sheet. Consistent with this change, the duration for both SBIR and STTR Phase I projects were adjusted to make them of equal duration, nine months each. With equal Phase I durations, the department can evaluate the Phase I and Phase II awards for both programs simultaneously, with significant increases in efficiency. (For Phase II, DoE uses the 2-year duration prescribed by the SBA.)
>
> **2000—Congress mandates SBIR study by the National Academy of Sciences.** In response to this mandate, DoE negotiated a contract for the study with the NAS and began to respond to NAS inquiries with respect to the study.
>
> **2005—Technology Niche Assessment (TNA) another assist program is added.** In addition to the Commercialization Assistance Program offered by Dawnbreaker, DoE introduced a new solicitation for innovative assistance program to help the SBIR firms to better commercialize technology. Foresight with their TNA program won the solicitation. Now, Dawnbreaker and Foresight both offer their programs to SBIR firms, and both are small businesses.

reassignment of responsibility and the manner in which it is carried out are described in section 5.4.2.

8.2 RESOURCES FOR PROGRAM ADMINISTRATION

Public Law 102-564 prohibits the federal agencies from using any of their SBIR budgets to fund the administrative costs of the program. As a result, DoE regards the administrative expenses as an additional, though unspecified, tax. As host for the SBIR office, DoE's Office of Science is responsible for the direct costs of administering the program: salaries for the federal employees, support services contracts, costs of developing and maintaining the electronic grant management system, etc.

Because SC must use its own funds to administer the SBIR program, and because of the historical resistance to SBIR at DoE, there has been a tendency for SC to reduce the resources for administering the program—ultimately to the point where 1.5 full time DoE employees, assisted by five contractors, were managing the administration (though not the review) of approximately 1,500 applications annually (both Phase I and Phase II). In March 2005, DoE SBIR reported a staff of three federal employees and six contractors. They still reported substantial resource constraints, including the use of 20-25 percent of their time on redundant paperwork issues.

In order to cope with these resource constraints, the SBIR staff has placed a high premium on sticking to a clearly mapped-out schedule and meeting established deadlines. These in turn are appreciated by TTMs and TPMs.

Management results over time—on time service, the cooperation of the technical program staff, acceptable outcomes—might seem to indicate that the program receives approximately the appropriate level of agency funding, even though the level of administrative support per dollar of funds awarded, or per application processed, is lower at DoE than at other study agencies.

Low funding levels for administration mean however that the DoE SBIR staff devotes nearly 100 percent of staff time to managing the processes for generating technical topics, and for receiving, evaluating, and selecting grant applications. This leaves little time for activities such as outreach, measuring Phase III activity, encouraging Phase III activity (both within and outside the department, including the national laboratories), internal evaluation, strategic planning, and documenting successes.

Reallocating resources toward SBIR administration in an agency whose culture has long favored more traditional research performers (national laboratories, universities) will be a challenge.

8.3 TOPIC GENERATION

Since at least 1992, SBIR topic development at DoE has proceeded as follows. In the spring, approximately six months prior to publication of the program solicitations in the fall, the DoE SBIR program manager sends a "call for topics" memo to all of the Department's SBIR portfolio managers (the designated representatives responsible for managing the SBIR portfolio within each of DoE technical program areas), along with copies to all DoE personnel that managed technical topics in the prior year. Accompanying the memo is the National Critical Technologies List,[4] as well as guidelines for topic development.[5] The SBIR portfolio managers develop their own specific procedures for generating the technical topics.

Once proposed topics are received from the portfolio managers, the DoE SBIR program manager arranges for a technical edit of the topic statements into a common format, beginning with an articulation of the problem to be addressed, often followed by a description of some technical approaches that are considered to be of interest. Prior to publication of the topic, the author of the topic provides the SBIR program manager with at least five bibliographic references (sometimes including Web sites), to be published with the topic description. In FY2003, this process resulted in 47 technical topics within 11 program areas.

All technical topics in the SBIR solicitation are constructed to support the overall mission of the technical program areas that provides the topics. Therefore, any proposal deemed to be responsive to the topic (in the First-Step Technical Review described below), and subsequently selected for award, would, by definition, be using DoE SBIR funds to support DoE mission. One hundred percent of DoE SBIR awards satisfy this condition.

DoE explicitly uses narrowly-defined technical topics to keep the number of applications at what the agency views to be a manageable level. Staff noted in interviews that increases in the number of applications may strain the resources available to provide a thorough evaluation. This is an important point. By defining topics narrowly—and by specifying in some cases the technical means to be used to address a topic—DoE is explicitly limiting the number of likely respondents, and is de facto excluding firms that might have innovative approaches, or may recognize problems that DoE has not yet become concerned about. Narrowing the topics is therefore a trade-off: It reduces agency work-load, but also agency opportunity.

A few of DoE case study companies indicated a preference for broader topics; others commented that tight topics gave the impression that the topic may be designed to give a particular company an advantage in the selection competition.

8.4 AWARD SELECTION

8.4.1 First-step Technical Review

The technical review of SBIR proposals takes place in a two-step process. In the first step, conducted by the Technical Topic Managers (TTMs), a proposal can be declined for any one of five reasons:

---

[4] As conveyed to DoE from the Small Business Administration.

[5] For example, Executive Order 13329, which encourages innovation in manufacturing. Access at <http://www.whitehouse.gov/news/releases/2004/02/20040224-6.html>.

1. It is deemed unresponsive to the solicitation topic and subtopic;
2. It is not for research or for research and development;
3. Not enough information is provided to properly evaluate the proposal;
4. It unduly duplicates other work; and
5. Its scientific and technical quality are deemed significantly lower than other proposals in the competition.

The last of these reasons requires the TTM to score the proposal with respect to the evaluation criteria. The objective of the first step of the review process is to reduce workload, and to avoid having reviewers waste time with non-responsive or poor proposals.

8.4.2 Initial Review Approaches

Until 1995, each Phase I proposal was assessed according to five criteria:

1. The scientific/technical quality of the proposed research;
2. The proposal's degree of innovation;
3. Staff qualifications and the availability of adequate facilities or instrumentation;
4. The anticipated benefits, technical and/or economic, of the proposed research (Phase I and Phase II), with special emphasis on the attraction of further funding;
5. The extent to which the Phase I award could prove the feasibility of the concept.

For each criterion, every application received a score from zero (unsatisfactory) to four (excellent). The TTMs (or TPMs, Technical Project Monitors, in Phase II) would receive the comments from expert technical reviewers, usually from outside DoE, and would then score the proposal based on these comments.

This approach resulted in tensions between the SBIR office and the technical programs. TTMs and TPMs raised multiple concerns, largely focused on the workload: Early SBIR solicitations attracted at least twice as many proposals, per dollar awarded, than other DoE programs. Also, the need to maintain program uniformity, ensure each applicant was treated fairly, guarantee confidentiality of proprietary information, and conform to SBA rules added extra steps and paperwork. Finally, DoE's commitment to on-time service[6] exacerbated demands on the TTMs and TPMs.

---

[6]From the very beginning, DoE SBIR program emphasized the need to meet its deadlines. The reasons for this were twofold: (1) the deadlines were mandated in the SBIR Policy Directive, issued by the Small Business Administration to guide agency operations, as stipulated in P.L. 97-219 (although most other agencies did not meet the SBA guidelines); and (2) even more importantly, DoE believed that a timely response to its applicants was only fair—after all, most had put in a lot of time on their proposals, did not deserve to be left in limbo, and were anxious to begin work.

PROGRAM MANAGEMENT   97

In short, technical program personnel believed that they were being asked to do an exorbitant amount of work for a small percentage of their budget, with research performed by a new type of performer with whom they had little experience and some doubt as to their capabilities. Technical program managers were also concerned that they were not getting a reasonable return on their SBIR investment.

8.4.3 1995 Process Revisions

In 1995 a Process Improvement Team was established to resolve these tensions (see Box 8-1 for program chronology). The team recommended changes to the grant selection process, which have evolved to the procedures that are used today.

The first-step review process and the selection of reviewers remained the same. The major changes concerned scoring and award selection.

First, the scoring system was made less rigid and more subjective. The number of criteria was reduced from five to three:

- Quality of the scientific/technical approach;
- Ability to carry out the project; and
- Impact.

Second, the primary authority for award selection was shifted to the technical programs. Each program's SBIR portfolio manager is responsible for developing a process for determining which grant applications should be selected for award.

While the SBIR program manager continues to oversee scoring, this process is now less formal. Within the SBIR office, funding candidates are reviewed to ensure that the scoring can be justified by the comments of the reviewers. If a discrepancy occurs, the SBIR program manager contacts the TTM to resolve the scoring. If the discrepancy cannot be resolved, the proposal is referred to an Adjudicating Official, appointed by the Director of the Office of Science.

The new grant selection process has worked well from this perspective. Each year, only one or two proposals are referred to the Adjudicating Official.

8.4.4 Fairness of Competition

The interviewed firms that commented on the fairness of competition (Eltron, NanoScience, NexTech) believed that the award system was fair. They appeared to understand that subjectivity plays a role in the selection of proposals for award, but this subjectivity was regarded as a general weakness of the peer review system and not a specific weakness of the SBIR award process.

One firm (Eltron) perceived an unevenness of competition between applicants that were academic off-shoots and those that were independent private companies. In a university setting, a grantee may have the advantage of hiring post-docs for little money or have access to university equipment at no extra charge. It was suggested that this potential disparity should be taken into consideration by decision makers.

8.5 OUTREACH

Among the consequences of limited administrative funding is the program's constrained ability to conduct outreach to the small business community and to others. Because the staff is so small and the demands of the evaluation processes are so large, DoE staff decline most speaking invitations. DoE does participate in the SBIR National Conferences, which have been held semi-annually and are sponsored by the DoD and the NSF. However, DoE has attempted to avoid state and local conferences because: (1) its staff was extremely small and busy with its own SBIR processes, (2) the National Conferences served the same purpose, and (3) too often, the audience was either too small or inappropriate (i.e., companies with no research capabilities) to justify the time and expense.

DoE does take steps to make small businesses aware of upcoming SBIR solicitations. Historically, DoE posted its solicitation in the *Commerce Business Daily* and mailed out twenty thousand copies each year in order to ensure that a broad community of small business was aware of DoE SBIR program.

The NRC SBIR Phase II Survey did seek information from firms regarding the extent to which they sought outside assistance in preparing their proposals from state agencies, mentor companies, regional associations, or universities. About 86 percent of respondents reported receiving no assistance in proposal preparation. A few firms received assistance from state agencies, or from mentor companies. Universities provided assistance to 8 percent of the respondents. Of those that received assistance, 60 percent found it very useful, 35 percent found it somewhat useful, and 5 percent did not find it useful.

8.6 THE APPLICATION AND AWARD PROCESS: AWARDEE COMMENTS

Half of the case study companies (Atlantia, Eltron, IPIX, NanoScience, Pearson) expressed overall satisfaction with the application process. Two of the three interviewees (IPIX, NanoScience) that commented on the feedback provided to applicants indicated that the comments were informative and straightforward. The third company (PPL) did not believe the feedback was particularly useful.

There were a number of recommendations with respect to the solicitations and the submission of proposals. Two case study companies (Airak, NexTech) recommended that all agencies should increase the issuance of their solicitations to at least twice per year, as DoD and NIH now offer. One of these interviewees (Airak) recommended a combined SBIR application process for all agencies, in order to save time and effort on behalf of the applicants.

One company indicated that the amount of work necessary to submit proposals outweighed the resulting funding, and another (Airak) was ambivalent about whether the preparation of SBIR proposals was an effective use of resources. However, four interviewees (Atlantia, Eltron, IPIX, NanoScience) believed that the cost and time spent are more than made up for by the benefits. One of these companies (NanoScience) stated that the learning process involved in putting together a strong application could be beneficial, sometimes more than any eventual reward.

Regarding topic specification, several interviewees (IPIX, NanoScience, PPL) were ambivalent about the need for broader topics; but where preferences were expressed (Atlantia, IPIX, PPL), they supported broader topics in order to increase transparency, ease topic selection for applying firms, and allow easier entry by smaller participants. (IPIX) One case study company "almost did not do a proposal" because its research did not appear to fit a published solicitation topic; yet, the work led to great success (Atlantia). Three interviewees (IPIX, NexTech, PPL) suspected that, occasionally, very tight topic specifications could imply a lock on the grant by a particular applicant whose orientation and capabilities are better reflected in the specifications.

8.7 MANAGING INFORMATION ON AWARDS

8.7.1 Reporting Requirements

DoE collects three types of information from SBIR participants:

- First, the SBIR proposals themselves. Data on all proposals for the last 20 years (successful and not) are maintained in a FoxPro database.
- Second are the traditional progress and final reports of the research project itself (final reports only for Phase I), which are delivered to both the TPM and the contracting officer. Final reports for Phase I funded projects that do not receive Phase II funding and Phase II final reports are delivered to the TPM and the department's Office of Scientific and Technical Information. The firms' reports are held for four years in confidentiality after completion of the SBIR effort, as required by P.L. 102-564.[7]
- Third, as a condition of their grants, Phase II awardees are required to report Phase III information for three years after their SBIR award and voluntarily thereafter; however, not all awardees actually do so.

8.7.2 Freedom of Information Act

One of the main purposes of the SBIR program, the commercialization of SBIR-developed technology, would be defeated if the results of the research were

---

[7]This law, P.L. 102-564, does state explicitly when this 4-year period begins.

made public. Therefore, the SBIR office has worked with DoE's Freedom of Information Office to identify legal exemptions that could be used to protect the companies' proprietary information both in their proposals and reports.

8.8 PROGRAM STRUCTURE

8.8.1 Differences Between Agencies

Most of the case study companies indicated an awareness of the differences between agencies in their administration of the program, but none indicated that they had a problem with the fact that there were differences. In fact, one company (Creare) called these differences "a strength" of the program. Two companies (Eltron, NexTech) understood that the administrative differences were intended to serve the agencies' differing missions and goals; for example, NSF procedures are consistent with their pursuit of more basic and long-term research, leading to commercial products, while DoD seeks solution-oriented proposals with more immediate military applications. Some specific differences between agencies will be identified within the categories that follow.

8.8.2 Award Limits

Three interviewees (Creare, IPIX, NanoScience) believed that the size of the awards is about right, although one of them (IPIX) preferred a slight increase for Phase II. In general these companies prefer the current situation to an environment in which there were fewer but larger grants. One interviewee commented that the current funding and award ratio allows for a wide variety of topics to be presented by the agencies while also encouraging firms to keep applying for SBIR grants. A large increase in funding might lead to complacency on the part of the applying firms, as they would be able to stretch the award over a longer period of time (IPIX).

Two others (Eltron, NexTech) advocated an increase in the award limits to keep up with inflation. The final interviewee that commented on the size of awards (Pearson) recommended a large increase for the Phase II limit, to $2 million.

8.8.3 Time Frames

The case study companies commented on three types of time frame: (1) the time that it takes to be paid following the submission of an invoice, (2) the time between Phase I and Phase II, and (3) the time for the entire process—from application submission to the end of Phase II—to be completed.

The first time frame received the most attention. Although one company (IPIX) said that funding delays, if existent, have been inconsequential, most of the others recognized differences between the agencies. DoE was considered to be the most prompt agency, with payments occurring fairly rapidly (Eltron). One interviewee (Atlantia) said that after receiving its SBIR awards, DoE sent checks for the amount of the awards and required little oversight. While switching to electronic filing and payments has sped up the process in some agencies, it was reported that others take longer to make payments; e.g., in the DoD.

The other time frames were cited once each by the interviewees. One case study company (IPIX) said that the time gap between the Phase I and Phase II requires attention, as it is difficult for some firms to maintain employees or facilities during the intervening time. Another (Airak) recommended that ways should be sought to speed up the total award time frame—four years from Phase I application to Phase II completion—including an option that would allow grantees to work more quickly, in order to aid future commercialization efforts.

8.8.4 Gaps Between SBIR Phase I and Phase II Funding

Another agency-wide issue for the SBIR program concerns the gap in funding between Phase I and Phase II. According to the NRC Phase II Survey, 36 percent of the responding projects from DoE experienced no funding gap between the end of Phase I and the beginning of the Phase II project targeted in the survey.

For the remaining firms, the funding gap affected work on the project: 56 percent stopped work on the project during the gap and another 33 percent continued to work during the gap but at a reduced pace. Three percent of firms reported ceasing all operations during the gap. Only 3 percent of the respondents received bridge funding of some kind during the gap. The average gap, as reported by 95 respondents, was 5 months. Three percent of respondents reported a gap of one or more years. The department's target for the Phase II gap remains three months.

During the early years of the program, DoE did have in place a gap-closing program, which allowed Phase I awardees to apply early for Phase II, thus effectively cutting the gap between Phases. However, this approach was eliminated as because the additional round of proposal reviews and scoring was seen as an added burden for the SBIR office and well as for the TTMs and TPMs in the technical program office.

8.9 PARTICIPATION OF DOE NATIONAL LABORATORIES IN SBIR

8.9.1 Overview of DoE National Laboratories

Among the federal agencies that participate in the SBIR a program, the Department of Energy is unique with respect to its government owned-contractor operated national laboratories (GOCOs). The uniqueness has to do with their large number and breadth of interests. While some other agencies have GOCOs (e.g., NASA's Jet Propulsion Laboratory and NSF's National Center for Atmospheric Research), they are relatively few in number. For the most part, the other SBIR agencies have government owned-government operated laboratories (GOGOs), for example the Department of Commerce's National Institute of Standards and Technology (NIST) or the Navy's Naval Research Laboratory (NRL). (DoE also has a GOGO, its National Energy Technology Laboratory.) GOCOs are not staffed by government employees, and, consequently, the work that they do is not being performed by the government. Therefore, GOCOs, including DoE national laboratories, are eligible to serve as subcontractors in SBIR since their participation would not violate the prohibition against SBIR dollars going back to the government.

The major national laboratories within DoE can first be distinguished by DoE programs they serve. The majority are either Science laboratories (i.e., "owned" by DoE's Office of Science (SC)) or Defense laboratories (i.e., "owned" by DoE's National Nuclear Security Administration, which administers the defense programs that are exempt from the SBIR set-asides). The Science laboratories can be multiprogram laboratories (serving a variety of technical areas within SC) or single-program laboratories.

The multiprogram Science laboratories and all three Defense laboratories are huge, with annual budgets on the order of hundreds of millions of dollars; some exceed $1 billion. Budgets for the SC single program laboratories are typically in the tens of millions, up to about $100 million. Together, these laboratories represent a major national resource.

The national laboratories also represent a resource for small businesses participating in SBIR. For the most part, their facilities and expertise cannot be duplicated elsewhere—in fact, the national laboratories are not allowed to compete with the private sector. As with universities, when a national laboratory is identified as a subcontractor on an SBIR proposal, the credibility of the research team is enhanced. Therefore, one might expect that a large number of small business/national laboratory partnerships would exist. However, this is not the case. Typically, less than 10 percent of DoE SBIR projects involve such partnerships. Many more projects have subcontracts with universities.

8.9.2 Why SBIR Collaborations Are Not More Frequent

The reasons for the infrequent utilization of the resources of the national laboratories may include:

(1) A lack of outreach on the part of most National Laboratories;
(2) Some trepidation among small companies in dealing with such large organizations, particularly when it comes to having to negotiate cooperative research and development agreements (CRADAs) or related vehicles;
(3) The fact that some of the other agencies do not allow their SBIR awardees to partner with DoE National Laboratories; and

(4) The recent SBA change in their policy directive requiring small businesses to apply for a waiver in order to partner with a DoE National Laboratory.[8]

The National Laboratories would appear to have an incentive to reach out to small businesses to encourage partnerships on SBIR projects.[9] After all, a significant part of the SBIR set asides represents money that would have passed from DoE to the laboratory, and collaboration on these projects would represent one means of recouping a portion of these funds. In addition, some laboratories have technology transfer goals that could be satisfied by these collaborations. Also, some laboratories, desiring to be good neighbors, seek to support the small businesses in their communities. However, the laboratories have been widely divergent with respect to their outreach efforts, ranging from aggressive for some to disinterested for others.

For reasons that are not entirely clear, however, the National Laboratories are treated differently in the two policy directives—for SBIR and STTR—promulgated by the Small Business Administration. For STTR (not covered by this NRC study) the enabling legislation (P.L. 102-564), along with the SBIR Policy Directive, specifically allows federally Funded Research and Development Centers (FFRDCs), which are identified by the National Science Foundation and includes DoE National Laboratories, to serve as the research institution (a required subcontractor on all STTR projects). However, for SBIR, the SBA Policy Directive imposes additional restrictions on small businesses that desire to partner with a National Laboratory: if the small business certifies that the work to be performed by the laboratory is important to the project and cannot be found elsewhere, the SBA will waive the restriction. DoE SBIR office assists its applicants in negotiating this waiver request process, and, to this point, no request has been denied by the SBA.

8.10 DEVELOPMENTS IN PROGRAM ADMINISTRATION SINCE 2003

8.10.1 Online Capabilities and Plans

The technology for administering grant applications at DoE was paper-based throughout the focal period of this study (1992-2003). In fall 2004, subsequent to the first meeting with the Academies' study team, the DoE initiated a pilot effort

---

[8]At DoE, the granting of such waivers is routine. SBIR program staff report that that the waiver process has been made minimally burdensome to small businesses. Nonetheless, for firms that may be unaware of the strong likelihood of being granted such a waiver, this additional requirement does create an additional barrier to collaboration with national laboratories.

[9]Of course, the laboratories must adhere to fairness-of-opportunity constraints, which, for example, would restrict a laboratory from soliciting a particular company as a partner for an SBIR project; however, laboratories may alert broad numbers of small businesses of their expertise with respect to technical topics in agency solicitations, or broadly invite small businesses to a laboratory open house.

to explore the possibility of transitioning to an electronic format for applications. Since 2005, all SBIR applications have been handled electronically. This adoption of a new procedure has imposed significant start-up costs in setting up the new technology and in the time required to process and manage applications. These costs are expected to diminish as the new system becomes increasingly familiar.

8.10.2 Program Manager Given More Control

In a recent change, DoE SBIR program managers are given more direct control over SBIR awards. With more control the program managers have taken more ownership and appear to be more engaged with the companies, resulting in positive impacts for everyone involved. Two specific areas of this change include the program managers deciding on technical assistance, and the program managers' sign-off on the second year of the Phase II awards.

- **Technical assistance**. The program managers now choose which of their award companies will be offered technical assistance support. Unfortunately, there was—and still remain—insufficient resources to fund free technical assistance for every award company. Until recently, support was awarded on a first-come first-served basis. With the recent change, all companies now have the opportunity to be selected by the program manager, whose judgment can be based on criteria such as the company's timeline and needs.
- **Year two sign-off**. DoE SBIR program managers now sign off on the second year of the Phase II awards. In the past, program managers had no control over funding once the award was granted. It was hard to terminate an award even if no progress had taken place. DoE expects that nearly all projects will receive the year two sign-off, but this additional capability gives the program manger more power and more accountability.

8.10.3 Phase II Supplemental Awards

In 2005, DoE began to offer firms completing Phase II awards the option of applying for a $250,000 supplemental award. This request was subject to review. The intention was to provide the highest performing firms with additional resources, as needed, to advance development of the technology in question.

However, complications in the administration of this supplemental program put this program on hold for some time. However, DoE SBIR staff reintroduced this option for funded firms in 2007, and it is now standard practice.

8.11 ACTIONS TAKEN BY DOE SBIR PROGRAM TO ENCOURAGE COMMERCIALIZATION

DoE has taken four distinct actions to encourage, promote, and track the commercialization of SBIR technology including:

1. Mandating evidence of commercialization within the Phase II evaluation criteria;
2. Providing commercialization assistance services to SBIR awardees;
3. Collecting Phase III data from Phase II awardees, and
4. Publicly recognizing success.

8.11.1 Evidence of Commercialization Included in Phase II Criteria

Public Law 102-564, which in 1992 both reauthorized the SBIR program and created the STTR program, also modified the definition of Phase II.[10] Essentially, this provision requires agencies to consider commercial potential according to four indicators in the evaluation of Phase II proposals, but it did not say how this should be done. At DoE, this requirement was implemented formally in the scoring of SBIR proposals. Three indicators were added to the criterion on Impact. These three indicators would count as one-half of the Impact criterion, or one-sixth of a proposal's total score. They are evaluated by the SBIR office, based on information that the small businesses are encouraged to include in their Phase II proposals.

DoE determined that the fourth indicator of commercial potential was already being addressed in the Impact criterion and was already being covered by reviewers in their comments on this criterion. Therefore, the fourth indicator was considered to be included within the other half of the score for the Impact criterion. All together, then, the four criteria count for somewhat more than one-sixth of the total score for a Phase II proposal.

8.11.2 Commercialization Assistance Services for SBIR Awardees

To aid Phase II awardees that seek to speed the commercialization of their SBIR technology, DoE has sponsored a Commercialization Assistance Program (CAP). The CAP provides, on a voluntary basis, individual assistance in develop-

---

[10]Public Law 102-564 states that "a second phase, to further develop proposals which meet particular program needs, in which awards shall be made based on the scientific and technical merit and feasibility of the proposals, as evidenced by the first phase, considering, among other things, the proposal's commercial potential, as evidenced by—(i) the small business concern's record of successfully commercializing SBIR or other research; (ii) the existence of second phase funding commitments from private sector or non-SBIR funding sources; (iii) the existence of third phase, follow-on commitments for the subject of the research; and (iv) the presence of other indicators of the commercial potential of the idea. . . ."

ing business plans and in preparing presentations to potential investment sponsors. The CAP is operated by a contractor, Dawnbreaker, Inc., a private firm based in Rochester, NY, which has repeatedly won competitions to provide this service.

The CAP is an intensive experience for the small businesses that choose to participate in it. The CAP begins with a kick-off meeting at DoE, where potential small business participants are introduced to the program. At that time they also meet one-on-one with Dawnbreaker staff to discuss their goals and technologies. Over the next several months, the companies, coached by Dawnbreaker staff, perform market research activities and prepare an iterative series of business plans. Because of the heavy workload (and for other reasons, e.g., illness), approximately 50 percent of the small businesses drop out during this period; however, this attrition is built into the design of the program.

The CAP culminates in a Commercialization Opportunity Forum, in which those companies that survive the business planning process present their business opportunities to a group of potential partners or investors—typically representatives from large corporations or from venture capital companies. Before the forum takes place, the companies are coached and rehearsed in the art of making effective presentations.

In order to make the forum a more attractive event for the potential partners/investors—i.e., to provide more business opportunities—DoE SBIR program often partners with other agencies or other DoE offices. These partners make their own contractual arrangements with Dawnbreaker for the training of their own small business participants. For the FY2006 Forum DoE SBIR agreed to partner with the National Science Foundation's SBIR Program and DoE's Office of Industrial Technologies. These partnerships will add another 20 and 10 SBIR Phase II projects, respectively, to the 30 DoE projects coming from DoE SBIR program, for a total of 60 business opportunities to be presented at the forum.[11]

Dawnbreaker, the DoE CAP vendor since 1989, tracks commercialization by polling each CAP participant at 6-, 12-, and 18-month intervals following the Commercialization Opportunity Forum, with emphasis on those companies that made a presentation at the forum. Dawnbreaker reported that half of the companies that completed the CAP program have already received in excess of $400 million for commercialization of their SBIR research. Dawnbreaker also solicits detailed feedback on CAP participation from clients through an evaluation template.

CAP was formally launched in 1989 by Program Manager Samuel Barish with a $50,000 contribution spread over 11 DoE technical programs.[12] At the present time, the CAP—along with the other commercialization assistance services described in this report—is funded from the 2.5 percent SBIR set-aside,

---

[11]Interview with Larry James, May 5, 2005.
[12]Interview with Sam Barish, DoE Director of Technology Research Division, May 6, 2005.

TABLE 8-1 2005 DoE SBIR Initiatives in Support of Commercialization

	CAP	Trailblazer	Technology Niche Assessment (TNA)	Virtual Deal Simulator™ (VDS)
Contractor	Dawnbreaker, Inc.	Foresight Science and Technology, Inc.	Foresight Science and Technology, Inc.	Foresight Science and Technology
Start Date	January 2005	January 2005	January 2005	January 2005
Completion Date	December 2007	December 2007	December 2007	December 2007
Eligibility	SBIR II only	SBIR I only	SBIR I or II	SBIR I or II

SOURCE: Department of Energy SBIR program publications and Web site.

in accordance with the SBA's 2002 *Final Policy Directive* to finance commercialization assistance, and supported by a 1997 finding by DoE legal counsel.[13] P.L. 102-564 permits the use of $4,000 per Phase I project for commercialization assistance.

In addition to the CAP, DoE now offers a number of other commercialization assistance services, which are summarized, along with the CAP, in Table 8-1.

DoE's 2005-2007 SBIR commercialization assistance "menu," which offers services to companies in SBIR Phases I and II, is funded at $2 million over three years. With a FY2005 SBIR budget of $102 million, DoE expects that approximately 180 SBIR winners from the 2004 pool of 425 Phase I and II recipients will participate in one or more of these components during 2005. (DoE expects some overlap, since many SBIR firms have multiple awards.) The agency's goal is successful commercialization of SBIR technology through licensing, subsidiary spin-off or "otherwise providing a path for the innovation to have the program manager's intended impact."[14]

One of these services, Technology Niche Assessments (TNA), has been available to DoE SBIR awardees for the past 6 years. In this service, which requires much less time commitment on behalf of the awardees, the companies initially describe their technology to a second DoE contractor, Foresight Science and Technology. Foresight then performs a search and analysis to identify private sector contacts that may have an interest as potential investors. Although the small businesses that have received this service have expressed general satisfaction, the SBIR office has not received any evidence that any small business received subsequent funding as a result of this service. However, TNAs performed during Phase I projects do allow companies to present more realistic commercialization plans in subsequent Phase II proposals.

---

[13]Interview with former DoE SBIR Program Manager Bob Berger, March 4, 2005.
[14]Interview with Larry James, December 29, 2004.

Over the history of DoE SBIR program, participation in commercial assistance services was open to all SBIR Phase II winners on a self-select basis, and attrition (i.e., dropping from the Dawnbreaker CAP) was the decision of the SBIR winner. With the addition of two new services on a pilot basis in 2005 (Trailblazer and Virtual Deal Simulator) DoE SBIR introduced a new participation model, which includes Phase I participants as well. The agency cites the importance of acknowledging commercialization tasks during Phase I.

Although all Phase I and II winners are encouraged to take advantage of one or more of the four services described, entry is determined primarily by personnel in DoE technical programs that support the SBIR program. The principal selection criterion is "technological innovation that has a high probability of success, and would have high program impact if successful." SBIR winners must be nominated by their project monitor, and must have won at least one DoE SBIR Phase I grant.

The agency anticipates 160 SBIR participants across the four commercial assistance services during 2005. Most recently DoE SBIR program has sought the involvement of TTM and TPMs in nominating for commercialization assistance firms whose technologies appear to have particular strong market potential.

The management of each commercialization assistance service is performed by its contractor, with regular SBIR client participation and fiscal reports made to the agency SBIR Program Manager. Included among the management responsibilities is tracking the commercialization progress of the SBIR participants. Each vendor reports on its SBIR clients' commercialization progress for a period of 2 years subsequent to completion of the Phase II grant. DoE SBIR uses both contractors' tracking results in biannual evaluations, but the data is not made publicly available—both vendors cite client confidentiality as the reason for not making commercialization figures available for each SBIR client.

8.11.3 Collecting Phase III Data

From the outset, DoE SBIR office made a concerted effort to assess SBIR commercialization outcomes.[15] All DoE SBIR Phase II awardees are required to report on their Phase III activity as a condition of their grant. Consequently, DoE has, on nearly an annual basis, collected Phase III data from its SBIR awardees.

---

[15]Outcomes are the link between broadly defined goals and the details of program management. Outcomes can be framed along three dimensions:

- Agency perspective: Did the program satisfy the objectives of the technical program managers? At the limit, were technical program managers ever willing to put "their own" money into the program?
- For the firms: Did the SBIR program help funded firms achieve goals including sustained employment for technical personnel, growth, and successful commercialization of new technologies?
- For society: Did SBIR funds create technologies that would not have been created otherwise? Did those technologies in turn motivate economic activity and improve human welfare?

The department learned, through trial and error, what they believe is the most effective way to ask awardees about Phase III. The key, they found, was to avoid asking about individual projects. A project-by-project approach was tempting because: (1) the agencies award and track individual projects, and (2) the GAO, in its earlier studies of SBIR, also used this approach. However, the project-by-project approach yielded inaccurate results because small businesses do not track their success in this way.

Small companies tend to track their success based on the products or services derived from SBIR projects. Therefore, DoE learned to ask companies to: (1) first, list all products and services that were derived from their DoE SBIR projects; (2) report on both sales and/or Phase III investment (including post-SBIR funding for further development) related to those products and services; and (3) then identify which Phase II projects contributed to the development of the products and services.

DoE SBIR office asks their Phase II awardees to report on two types of revenues: (1) sales, and (2) follow-on investment. Both categories are further broken down into federal and nonfederal sources of revenue. The second category, follow-on investment, could include funding for activities that are directly related to commercialization (marketing, setting up a production facility, etc.) or funding for further development of the technology. DoE considers all of these categories to be examples of "Phase III" funding.

Of the 787 companies that received 1,731 Phase II awards through 2002, 609 responded to DoE Phase II survey. Their responses are detailed in Table 8-2.

TABLE 8-2 DoE Sales and Development Outcomes

	Sales		
	FED	Nonfederal	Total
Companies	136	246	269
> $850,000	42	90	109
Amounts ($)	232,996,373	1,384,990,095	1,617,986,468
> $850,000	215,563,248	1,352,142,325	

	Development			
	Internal	Nonfederal	Fed Non-SBIR	Total
Companies	346	246	160	419
> $850,000	53	67	68	146
Amounts ($)	246,885,194	631,295,607	455,757,462	1,333,938,263
> $850,000	187,257,499	588,425,378	427,494,830	

SOURCE: Department of Energy.

Forty-eight percent of the respondents reported some sales, with 18 percent reporting sales of over $850,000 (a threshold value chosen as the total public funds awarded in a Phase I and Phase II, combined).[16]

Although many companies received Phase III funding, little or none of this funding came from DoE itself. A longtime DoE SBIR program manager could not recall one instance in which DoE has funded Phase III with non-SBIR funds as a follow-on to the research performed in Phases I and II.[17] DoE SBIR staff report that, during the interval covered by the study, many DoE Technical Project Monitors (TPMs) were unaware that SBIR technologies can be procured through a noncompetitive process in Phase III.

One reason for this information gap was that no systematic process existed for continually disseminating this information to DoE technical programs. On about 10-15 occasions, the aforementioned program manager provided DoE technical program managers with documentation to show that the law and the SBA Policy Directive allowed, even encouraged, agencies to make Phase III awards to the SBIR company on a sole source basis. However, he was not aware that such follow-on funding ever resulted from these efforts. (Phase III funding from DoE can occur without the knowledge of the SBIR office; for example, Phase III funding from national laboratories or DoE offices could be administered directly by the technical programs, without informing the SBIR office.)

8.11.4 Recognizing Success

DoE recognizes the success of SBIR firms in a number of ways. The agency maintains and updates a broad selection of commercially successful SBIR-funded technologies on its SBIR Web site (<http://www.science.DoE.gov/sbir>). The Web site also identifies DoE SBIR awardees who have received the prestigious R&D 100 Award.

---

[16]No data exist (self-reported or other) that would inform us regarding the extent to which SBIR funds were critical in the development of the products in question. Furthermore, it is possible that firms have an incentive to inflate these figures, and no obvious incentive to underreport "success" in commercialization in a survey of this variety. Although It is also possible that the lack of perfect institutional memory in small, dynamic firms could lead to systematic underreporting.

[17]The norm in the fossil energy and environmental offices is to work with very large companies, mostly on a cost-sharing basis. SBIR projects involving the development of instrumentation may an exception. One company reported a number of instances in which DoE Office of Health and Environmental Research (OHER, now the Office of Biological and Environment Research) provided Phase III funding for the use of instruments developed with SBIR funds in environmental fields.

Appendixes

Appendix A

DoE SBIR Program Data

TABLE APP-A-1 Phase I Awards

Fiscal Year	Number of Phase I Awards	Average Award Size ($)	Maximum Award Size ($)	Total Phase I Award Dollars
1992	196	49,765	50,000	9,753,886
1993	167	74,076	75,000	12,370,630
1994	209	74,363	75,000	15,541,961
1995	196	74,570	75,000	14,615,796
1996	167	74,795	75,000	12,490,776
1997	194	74,099	75,000	14,375,212
1998	204	74,669	75,000	15,232,440
1999	185	99,432	99,999	18,394,937
2000	292	68,152	99,998	19,900,505
2001	310	68,166	99,999	21,131,432
2002	328	68,894	99,999	22,597,310
2003	323	67,876	99,999	21,923,938
2004	257	95,325	99,999	24,498,613
2005	258	99,449	99,999	25,657,718
Total	3,286			248,485,154

SOURCE: U.S. Small Business Administration, Tech-Net Database.

TABLE APP-A-2 Phase I Awards to Minority-Owned Businesses

Fiscal Year	Number of Phase I Awards	Average Award Size ($)	Maximum Award Size ($)	Total Phase I Award Dollars
1992	20	50,000	50,000	1,000,000
1993	21	74,440	75,000	1,563,235
1994	23	74,919	75,000	1,723,137
1995	23	74,917	75,000	1,723,093
1996	23	74,800	75,000	1,720,398
1997	33	74,621	75,000	2,462,480
1998	34	74,723	75,000	2,540,584
1999	32	99,019	99,991	3,168,602
2000	46	62,888	99,997	2,892,843
2001	39	74,216	99,996	2,894,424
2002	37	62,040	99,988	2,295,495
2003	34	85,256	99,999	2,898,695
2004	31	95,535	99,997	2,961,589
2005	39	99,954	99,999	3,898,224

SOURCE: U.S. Small Business Administration, Tech-Net Database.

TABLE APP-A-3 Phase I Awards to Woman-Owned Businesses

Fiscal Year	Number of Phase I Awards	Average Award Size ($)	Maximum Award Size ($)	Total Phase I Award Dollars
1992	8	50,000	50,000	400,000
1993	17	73,621	75,000	1,251,555
1994	18	74,897	75,000	1,348,142
1995	5	70,964	75,000	354,818
1996	9	74,777	75,000	672,991
1997	18	74,644	75,000	1,343,594
1998	14	74,925	75,000	1,048,951
1999	19	98,671	99,997	1,874,754
2000	24	70,803	99,997	1,699,276
2001	25	67,989	99,999	1,699,728
2002	18	66,240	99,988	1,192,327
2003	25	75,144	99,997	1,878,605
2004	19	98,074	99,999	1,863,400
2005	33	99,578	99,998	3,286,065

SOURCE: U.S. Small Business Administration, Tech-Net Database.

TABLE APP-A-4 Phase I Awards, by State

State	Number of Phase I Awards	State	Number of Phase I Awards
AL	20	CA	704
AR	8	MA	569
AZ	65	CO	311
CA	704	CT	185
CO	311	NY	123
CT	185	TX	121
DC	1	NJ	113
DE	32	VA	113
FL	42	OH	97
GA	22	WA	83
HI	1	NM	82
IA	3	PA	72
IL	64	MD	69
IN	16	AZ	65
KS	7	IL	64
KY	4	TN	64
LA	3	UT	55
MA	569	FL	42
MD	69	MI	41
ME	7	DE	32
MI	41	OR	32
MN	25	NC	27
MO	6	MN	25
MS	1	GA	22
MT	11	NH	22
NC	27	AL	20
ND	3	WI	17
NE	3	IN	16
NH	22	MT	11
NJ	113	OK	11
NM	82	AR	8
NV	8	NV	8
NY	123	KS	7
OH	97	ME	7
OK	11	VT	7
OR	32	MO	6
PA	72	WV	6
RI	2	WY	5
SC	3	KY	4
TN	64	IA	3
TX	121	LA	3
UT	55	ND	3
VA	113	NE	3
VT	7	SC	3
WA	83	RI	2
WI	17	DC	1
WV	6	HI	1
WY	5	MS	1
		Total	3,286

SOURCE: U.S. Small Business Administration, Tech-Net Database.

TABLE APP-A-5 Phase I Awards in Massachusetts, by Town/City

City	Number of Phase I Awards		
Waltham	81	Summary	
Watertown	60	48 towns win 1 award	
Bedford	52	Top 5 towns	278
Andover	46	Top 5 %	45.6%
Shrewsbury	39		
Somerville	33		
Westborough	28		
Burlington	27		
Norwood	23		
Cambridge	20		
Billerica	19		
Woburn	19		
Turners Falls	18		
Marlborough	15		
Natick	12		
Lowell	11		
Newton	10		
Boston	9		
East Falmouth	8		
Lexington	8		
Concord	6		
Littleton	6		
Chelmsford	5		
Sturbridge	5		
Wilmington	5		
Amherst	4		
Belmont	4		
Topsfield	4		
Duxbury	3		
Holliston	3		
Needham Heights	3		
Willmington	3		
Allston	2		
Forestdale	2		
Salem	2		
Wayland	2		
Acton	1		
Boxboro	1		
Brookline	1		
Chatham	1		
Danvers	1		
Dedham	1		
North Chelmsford	1		
North Falmouth	1		
South Deerfield	1		
Tewksbury	1		
Wakefield	1		
Wellesley	1		
Total	609		

SOURCE: Department of Energy.

TABLE APP-A-6 Phase I Awards, by Company

Company	Number of Phase I Awards	Total Dollars
Physical Optics Corporation	45	3,824,563
TDA Research, Inc.	40	3,249,287
Omega-p, Inc.	39	3,449,995
Radiation Monitoring Devices, Inc.	39	3,350,000
Eltron Research, Inc.	36	2,974,855
Physical Sciences, Inc.	34	2,723,934
Membrane Technology and Research, Inc.	32	2,599,619
Supercon, Inc.	30	2,474,763
ADA Technologies, Inc.	27	2,273,683
MER Corporation	25	2,074,875
Ceramem Corporation	25	2,225,000
Tech-x Corporation	23	2,198,739
Fm Technologies, Inc.	21	1,649,531
Hypres, Inc.	20	1,598,387
Advanced Fuel Research, Inc.	20	1,522,906
Science Research Laboratory, Inc.	19	1,498,107
American Superconductor Corporation	19	1,550,000
Lynntech, Inc.	18	1,566,346
Spire Corporation	17	1,241,391
Calabazas Creek Research	17	1,575,000
Top 20 companies total	546	45,620,981
All	3,286	69,069,709

SOURCE: U.S. Small Business Administration, Tech-Net Database.

TABLE APP-A-7 Phase I Awards, by Zip Code

Zip code	Number of Phase I Awards	Summary	
80301	103	Top 20	965
90501	72	Next 30	575
80033	64	Next 50	515
92121	63	All others	1,410
06520	56		
02472	55		
02453	53		
01730	52		
94025	49		
80127	48		
01810	46		
85706	41		
01545	39		
95054	34		
20151	34		
02143	33		
06108	32		
10523	32		
77840	30		
95070	29		
90275	28		
01581	28		
01803	27		
91324	27		
37830	24		
02062	23		
06704	22		
97701	21		
90505	21		
87109	21		
80501	20		
06813	20		
80026	19		
01821	19		
01801	19		
02451	18		
01376	18		
48108	17		
94086	16		
08852	16		
30341	16		
03755	16		
87505	16		
06810	15		
06473	15		

SOURCE: Department of Energy.

TABLE APP-A-8 Phase II Awards

Fiscal Year	Number of Phase II Awards	Average Award Size ($)	Maximum Award Size ($)	Total Phase II Award Dollars
1992	66	488,010	500,000	32,208,670
1993	72	498,793	500,000	35,913,116
1994	61	596,332	600,000	36,376,260
1995	77	725,065	750,000	55,829,971
1996	70	734,996	750,000	51,449,732
1997	82	722,697	750,000	59,261,141
1998	83	741,516	750,000	61,545,841
1999	85	696,165	750,000	59,174,004
2000	91	711,715	750,000	64,766,071
2001	98	676,185	900,000	66,266,094
2002	103	694,454	750,000	71,528,733
2003	103	706,382	892,342	72,757,388
2004	115	719,961	750,000	82,795,485
2005	107	699,557	750,060	74,852,565
Total	1,213			824,725,071

SOURCE: U.S. Small Business Administration, Tech-Net Database.

TABLE APP-A-9 Phase II Awards to Minority-Owned Businesses

Fiscal Year	Number of Phase II Awards	Average Award Size ($)	Maximum Award Size ($)	Total Phase II Award Dollars
1992	11	478,175	500,000	5,259,929
1993	8	495,534	500,000	3,964,271
1994	9	599,976	600,000	5,399,783
1995	10	727,726	750,000	7,277,262
1996	8	745,948	750,000	5,967,580
1997	7	692,941	750,000	4,850,585
1998	11	749,629	750,000	8,245,924
1999	12	688,043	750,000	8,256,518
2000	17	701,433	750,000	11,924,363
2001	10	698,000	900,000	6,980,004
2002	14	709,941	750,000	9,939,167
2003	5	649,840	749,996	3,249,199
2004	12	733,284	750,000	8,799,409
2005	13	709,913	750,000	9,228,865

SOURCE: U.S. Small Business Administration, Tech-Net Database.

TABLE APP-A-10 Phase II Awards to Woman-Owned Businesses

Fiscal Year	Number of Phase II Awards	Average Award Size ($)	Maximum Award Size ($)	Total Phase II Award Dollars
1992	2	497,229	498,766	994,457
1993	1	500,000	500,000	500,000
1994	6	599,910	600,000	3,599,461
1995	9	749,966	750,000	6,749,696
1996	1	749,994	749,994	749,994
1997	4	712,408	750,000	2,849,630
1998	5	709,728	750,000	3,548,641
1999	6	687,239	750,000	4,123,432
2000	7	674,999	750,000	4,724,995
2001	8	712,377	900,000	5,699,018
2002	6	716,221	750,000	4,297,324
2003	6	739,434	892,342	4,436,602
2004	11	740,873	750,000	8,149,604
2005	6	683,979	749,868	4,103,872

SOURCE: U.S. Small Business Administration, Tech-Net Database.

TABLE APP-A-11 Phase II Average Awards, by Demographic

Fiscal Year	Average Award Size ($)		
	All Awards	Minority	Women
1992	488,010	478,175	497,229
1993	498,793	495,534	500,000
1994	596,332	599,976	599,910
1995	725,065	727,726	749,966
1996	734,996	745,948	749,994
1997	722,697	692,941	712,408
1998	741,516	749,629	709,728
1999	696,165	688,043	687,239
2000	711,715	701,433	674,999
2001	676,185	698,000	712,377
2002	694,454	709,941	716,221
2003	706,382	649,840	739,434
2004	719,961	733,284	740,873
2005	699,557	709,913	683,979

SOURCE: U.S. Small Business Administration, Tech-Net Database.

TABLE APP-A-12 Phase II Awards, by State

State	Number of Phase II Awards	State	Number of Phase II Awards
AL	9	CA	287
AR	3	MA	199
AZ	25	CO	116
CA	287	CT	74
CO	116	VA	45
CT	74	NJ	42
DC	0	NY	42
DE	10	OH	35
FL	15	TX	35
GA	6	WA	35
HI	0	NM	30
IA	0	PA	27
IL	18	TN	27
IN	4	AZ	25
KS	3	MD	24
KY	2	OR	19
LA	0	IL	18
MA	199	UT	16
MD	24	FL	15
ME	2	MI	14
MI	14	DE	10
MN	8	NC	10
MO	2	NH	10
MS	0	AL	9
MT	4	MN	8
NC	10	GA	6
ND	0	WI	6
NE	0	IN	4
NH	10	MT	4
NJ	42	AR	3
NM	30	KS	3
NV	1	WV	3
NY	42	KY	2
OH	35	ME	2
OK	2	MO	2
OR	19	OK	2
PA	27	NV	1
RI	1	RI	1
SC	1	SC	1
TN	27	VT	1
TX	35	DC	0
UT	16	HI	0
VA	45	IA	0
VT	1	LA	0
WA	35	MS	0
WI	6	ND	0
WV	3	NE	0
WY	0	WY	0
		Total	1,213

SOURCE: U.S. Small Business Administration, Tech-Net Database.

TABLE APP-A-13 Top Phase II Award Winners (Plus Ties)

Company	Number of Phase II Awards	Phase II Dollars Awarded
Physical Optics Corporation	21	14,399,812
TDA Research, Inc.	21	14,850,271
Omega-p, Inc.	18	12,650,000
Eltron Research, Inc.	18	12,099,295
MER Corporation	15	10,019,266
Radiation Monitoring Devices, Inc.	14	10,349,999
Membrane Technology and Research, Inc.	14	9,923,405
Fm Technologies, Inc.	13	9,099,404
Ceramem Corporation	13	9,200,000
Science Research Laboratory, Inc.	12	7,949,428
Calabazas Creek Research	12	7,395,865
Physical Sciences, Inc.	12	7,966,660
Tech-x Corporation	11	7,843,709
Duly Research, Inc.	11	7,699,508
Hypres, Inc.	10	6,599,415
Diversified Technologies, Inc.	10	6,436,557
American Superconductor Corporation	9	6,600,000
ADA Technologies, Inc.	9	6,299,009
Haimson Research Corporation	9	6,079,745
Bend Research, Inc.	8	4,184,884
Lynntech, Inc.	8	5,824,154
Fuelcell Energy, Inc.	8	5,249,873
Advanced Fuel Research, Inc.	8	5,136,686
Igc Advanced Superconductors, Inc.	8	4,849,929
Supercon, Inc.	8	4,599,731
Spire Corporation	8	5,494,425
Total	308	208,801,030
All awards	1,213	824,725,071
Top 20 Percentage of total	25.4	25.3

SOURCE: U.S. Small Business Administration, Tech-Net Database.

TABLE APP-A-14 Phase I Applications

Fiscal Years	Number of Declines	Number of Awards	Total Number of Phase I Applications	Phase I Success Rate (%)
1992	1,330	198	1,529	12.9
1993	2,296	200	2,495	8.0
1994	2,060	212	2,272	9.3
1995	1,369	199	1,568	12.7
1996	1,257	173	1,429	12.1
1997	1,021	199	1,220	16.3
1998	986	206	1,190	17.3
1999	933	203	1,133	17.9
2000	870	202	1,081	18.7
2001	674	213	897	23.7
2002	728	228	966	23.6
2003	996	219	1,223	17.9
Total	14,520	2,452	17,003	

SOURCE: Department of Energy.

TABLE APP-A-15 Phase I Applications—Woman-Owned Businesses

Fiscal Year	Agency Data Responses			Total Applications	Share of Applications (%)
	Yes	Blank	No		
1992	144	1,385		1,529	9.4
1993	293	2,202		2,495	11.7
1994	271	2,001		2,272	11.9
1995	166	1,401		1,567	10.6
1996	148	1,131	150	1,429	10.4
1997	158	739	323	1,220	13.0
1998	132	44	1,014	1,190	11.1
1999	94	26	1,013	1,133	8.3
2000	123	48	910	1,081	11.4
2001	69	21	807	897	7.7
2002	68	34	864	966	7.0
2003	114	28	1,081	1,223	9.3
Total	1,780	9,060	6,162	17,002	

SOURCE: Department of Energy.

TABLE APP-A-16 Phase I Applications—Minority-Owned Businesses

Fiscal Year	Agency Data Responses			Number of Awards	Total Applications	Share of Applications (%)
	Yes	No	Blank			
1992	442		2,053	20	2,515	17.6
1993	370		1,901	23	2,294	16.1
1994	235		1,333	23	1,591	14.8
1995	241		1,288	24	1,553	15.5
1996	223	704	293	26	1,246	17.9
1997	233	1,047	149	34	1,463	15.9
1998	193	843	45	34	1,115	17.3
1999	194	951	45	33	1,223	15.9
2000	151	784	31	30	996	15.2
2001	223	971	29	28	1,251	17.8
2002	185	923	25	22	1,155	16.0
2003	162	714	21	27	924	17.5
Total	2,852	6,937	7,213	324	17,326	16.5

SOURCE: Department of Energy.

TABLE APP-A-17 Phase I Applications—Success Rates by Demographics

Fiscal Year	Number of Declines				Number of Awards				Total Number of Applications			
	Woman	Minority	All Others	All	Woman	Minority	All Others	All	Woman	Minority	All Others	All
1992	144	442	943	1,529	13	20	165	198	157	462	1,108	1,727
1993	293	370	1,832	2,495	17	23	160	200	310	393	1,992	2,695
1994	271	235	1,766	2,272	18	23	171	212	289	258	1,937	2,484
1995	166	241	1,161	1,568	12	24	163	199	178	265	1,324	1,767
1996	148	223	1,058	1,429	13	26	134	173	161	249	1,192	1,602
1997	158	233	829	1,220	21	34	144	199	179	267	973	1,419
1998	132	193	865	1,190	14	34	158	206	146	227	1,023	1,396
1999	94	194	845	1,133	23	33	147	203	117	227	992	1,336
2000	123	151	807	1,081	17	30	155	202	140	181	962	1,283
2001	69	223	605	897	13	28	172	213	82	251	777	1,110
2002	68	185	713	966	12	22	194	228	80	207	907	1,194
2003	114	162	947	1,223	18	27	174	219	132	189	1,121	1,442
Total	1,780	2,852	12,371	17,003	191	324	1,937	2,452	1,971	3,176	14,308	19,455

Fiscal Year	Application Shares (%)				Success Rates (%)			
	Woman	Minority	All Others	All	Woman	Minority	All Others	All
1992	9.1	26.8	64.2	100	9.0	4.5	17.5	12.9
1993	11.5	14.6	73.9	100	5.8	6.2	8.7	8.0
1994	11.6	10.4	78.0	100	6.6	9.8	9.7	9.3
1995	10.1	15.0	74.9	100	7.2	10.0	14.0	12.7
1996	10.0	15.5	74.4	100	8.8	11.7	12.7	12.1
1997	12.6	18.8	68.6	100	13.3	14.6	17.4	16.3
1998	10.5	16.3	73.3	100	10.6	17.6	18.3	17.3
1999	8.8	17.0	74.3	100	24.5	17.0	17.4	17.9
2000	10.9	14.1	75.0	100	13.8	19.9	19.2	18.7
2001	7.4	22.6	70.0	100	18.8	12.6	28.4	23.7
2002	6.7	17.3	76.0	100	17.6	11.9	27.2	23.6
2003	9.2	13.1	77.7	100	15.8	16.7	18.4	17.9
Average					12.7	12.7	17.4	15.9

SOURCE: Department of Energy.

TABLE APP-A-18 Phase I Applications, by State

Alphabetical					Sorted by Number of Applications			
State	Number of Applications	Number of Awards	Success Rate (%)		State	Number of Applications	Number of Awards	Success Rate (%)
AK	7		0.0		CA	3,052	516	16.9
AL	181	15	8.3		MA	2,182	440	20.2
AR	82	7	8.5		CO	1,157	217	18.8
AZ	473	46	9.7		TX	788	80	10.2
CA	3,052	516	16.9		VA	776	78	10.1
CO	1,157	217	18.8		CT	694	150	21.6
CT	694	150	21.6		NY	636	90	14.2
DC	17		0.0		OH	604	71	11.8
DE	133	21	15.8		NJ	588	82	13.9
EC	1		0.0		PA	540	55	10.2
FL	361	25	6.9		NM	512	72	14.1
GA	160	21	13.1		MD	506	49	9.7
HI	18	1	5.6		AZ	473	46	9.7
IA	60	3	5.0		IL	471	46	9.8
ID	50		0.0		FL	361	25	6.9
IL	471	46	9.8		WA	343	55	16.0
IN	80	12	15.0		UT	333	41	12.3
IO	1		0.0		TN	324	53	16.4
KS	43	4	9.3		MI	253	35	13.8
KY	41	5	12.2		AL	181	15	8.3
LA	72	3	4.2		OR	178	33	18.5
MA	2,182	440	20.2		MN	166	20	12.0
MD	506	49	9.7		GA	160	21	13.1
ME	62	7	11.3		NC	151	23	15.2
MI	253	35	13.8		OK	145	8	5.5
MN	166	20	12.0		NH	140	21	15.0

127

State	N	%
MO	75	2.7
MS	23	4.3
MT	89	10.1
NC	151	15.2
ND	36	8.3
NE	9	11.1
NH	140	15.0
NJ	588	13.9
NM	512	14.1
NV	45	11.1
NY	636	14.2
OH	604	11.8
OK	145	5.5
OR	178	18.5
PA	540	10.2
PR	18	0.0
RI	26	3.8
SC	49	4.1
SD	21	0.0
TN	324	16.4
TX	788	10.2
UT	333	12.3
VA	776	10.1
VT	41	7.3
WA	343	16.0
WI	100	11.0
WV	37	13.5
WY	50	6.0
Other	3	33.3
Total	17,003	

State	N	n	%
DE	133	21	15.8
WI	100	11	11.0
MT	89	9	10.1
AR	82	7	8.5
IN	80	12	15.0
MO	75	2	2.7
LA	72	3	4.2
ME	62	7	11.3
IA	60	3	5.0
WY	50	3	6.0
ID	50		0.0
SC	49	2	4.1
NV	45	5	11.1
KS	43	4	9.3
KY	41	5	12.2
VT	41	3	7.3
WV	37	5	13.5
ND	36	3	8.3
RI	26	1	3.8
MS	23	1	4.3
SD	21		0.0
HI	18	1	5.6
PR	18		0.0
DC	17		0.0
NE	9	1	11.1
AK	7		0.0
EC	1		0.0
IO	1		0.0
All			9.7

SOURCE: Department of Energy.

TABLE APP-A-19 Phase I Applications and Success Rates, by Company Top 20 (Plus Ties)

Company	Number of Applications	Number of Awards	Success Rate (%)	
Eltron Research, Inc.	297	36	12.1	
Physical Optics Corporation, Electro-optics & Holo	220	45	20.5	
MER Corporation	218	25	11.5	
Lynntech, Inc.	150	18	12.0	
Advanced Fuel Research, Inc.	142	20	14.1	
Physical Sciences, Inc.	137	34	24.8	
TDA Research, Inc.	125	40	32.0	
Spire Corporation	124	17	13.7	
Radiation Monitoring Devices, Inc.	94	39	41.5	
American Superconductor Corporation	90	19	21.1	
Materials Modification, Inc.	89	1	1.1	*
Membrane Technology and Research, Inc.	89	32	36.0	
Ceramem Corporation	87	25	28.7	
Foster-Miller, Inc.	87	14	16.1	
Nanomaterials Research Corporation	87	14	16.1	
ADA Technologies, Inc.	83	27	32.5	
Technology International Incorporated	82	0	0.0	**
Tpl, Inc.	80	13	16.3	
Supercon, Inc.	78	30	38.5	
Ultramet	76	11	14.5	
Average			20.1	

Summary	
Top 20 average	20.1%
Without * and **	20.3%
All applicants	12.6%

SOURCE: Department of Energy.
Notes: (*) Technology International Inc. may not be a high level research company; (**) Materials Modification Inc. is a very high level research company, with many federal and commercial customers, plus R&D awards.

TABLE APP-A-20 Phase II Applications

Fiscal Year	Number of Applications	Number of Awards	Success Rate (%)
1992	159	66	41.5
1993	180	70	38.9
1994	183	72	39.3
1995	192	81	42.2
1996	171	72	42.1
1997	150	85	56.7
1998	174	85	48.9
1999	179	88	49.2
2000	184	91	49.5
2001	178	98	55.1
2002	189	103	54.5
2003	207	102	49.3
Total	2,146		

SOURCE: Department of Energy.

TABLE APP-A-21 Phase II Applications from Woman-Owned Companies

Fiscal Year	Agency Data Responses			All Others	All	Percent Share of Total
	Yes	Blank	No			
1995	17	175		175	192	8.9
1996	11	160		160	171	6.4
1997	12	13	125	138	150	8.0
1998	19	41	114	155	174	10.9
1999	12	6	161	167	179	6.7
2000	22	6	156	162	184	12.0
2001	12	8	158	166	178	6.7
2002	10	4	175	179	189	5.3
2003	12	8	187	195	207	5.8

SOURCE: Department of Energy.

TABLE APP-A-22 Phase II Applications—Minority-Owned Businesses

Fiscal Year	Agency Data Responses			All Others	All	Percent of All Applications
	Yes	Blank	No			
1995	21	170		170	191	11.0
1996	23	148		148	171	13.5
1997	21	12	117	129	150	14.0
1998	30	37	107	144	174	17.2
1999	27	6	146	152	179	15.1
2000	29	7	148	155	184	15.8
2001	26	8	144	152	178	14.6
2002	23	4	162	166	189	12.2
2003	22	8	177	185	207	10.6

SOURCE: Department of Energy.

TABLE APP-A-23 Phase II Applications, by State

Ordered Alphabetically					Ordered by Applications			
State	Number of Applications	Number of Awards	Success Rate (%)		State	Number of Applications	Awards	Success Rate (%)
AL	14	7	50.0		CA	472	22.0%	53.6
AR	6	2	33.3		MA	386	18.0%	43.8
AZ	41	23	56.1		CO	190	8.9%	47.9
CA	472	253	53.6		CT	131	6.1%	51.9
CO	190	91	47.9		NY	76	3.5%	47.4
CT	131	68	51.9		NJ	66	3.1%	48.5
DE	18	7	38.9		NM	64	3.0%	39.1
FL	24	11	45.8		OH	63	2.9%	49.2
GA	18	6	33.3		TX	60	2.8%	41.7
HI	1	0	0.0		VA	57	2.7%	61.4
IA	3	0	0.0		TN	55	2.6%	47.3
IL	34	11	32.4		PA	48	2.2%	35.4
IN	11	3	27.3		WA	48	2.2%	47.9
KS	2	1	50.0		AZ	41	1.9%	56.1
KY	2	1	50.0		MD	41	1.9%	51.2
LA	3	0	0.0		UT	39	1.8%	33.3
MA	386	169	43.8		IL	34	1.6%	32.4
MD	41	21	51.2		MI	33	1.5%	39.4
ME	6	2	33.3		OR	32	1.5%	53.1
MI	33	13	39.4		FL	24	1.1%	45.8
MN	18	10	55.6		NC	20	0.9%	55.0
MO	1	0	0.0		NH	20	0.9%	0.0
MS	1	0	0.0		DE	18	0.8%	38.9
MT	5	3	60.0		GA	18	0.8%	33.3
NC	20	11	55.0		MN	18	0.8%	55.6
ND	3	0	0.0		AL	14	0.7%	50.0

NE	1	0	0.0	IN	11	0.5%	27.3
NH	20	0	0.0	WI	10	0.5%	60.0
NJ	66	32	48.5	OK	7	0.3%	14.3
NM	64	25	39.1	AR	6	0.3%	33.3
NV	6	1	16.7	ME	6	0.3%	33.3
NY	76	36	47.4	NV	6	0.3%	16.7
OH	63	31	49.2	MT	5	0.2%	60.0
OK	7	1	14.3	WV	5	0.2%	60.0
OR	32	17	53.1	IA	3	0.1%	0.0
PA	48	17	35.4	LA	3	0.1%	0.0
SC	2	1	50.0	ND	3	0.1%	0.0
TN	55	26	47.3	KS	2	0.1%	50.0
TX	60	25	41.7	KY	2	0.1%	50.0
UT	39	13	33.3	SC	2	0.1%	50.0
VA	57	35	61.4	WY	2	0.1%	0.0
VT	1		0.0	HI	1	0.0%	0.0
WA	48	23	47.9	MO	1	0.0%	0.0
WI	10	6	60.0	MS	1	0.0%	0.0
WV	5	3	60.0	NE	1	0.0%	0.0
WY	2		0.0	VT	1	0.0%	0.0
Total	2,146	1,005	46.8		2,146		

SOURCE: Department of Energy.

TABLE APP-A-24 Phase II Applications—Success Rates by Demographics

Fiscal Year	Number of Declines				Number of Awards				Total Number of Applications			
	Woman	Minority	All Others	All	Woman	Minority	All Others	All	Woman	Minority	All others	All
1995	17	21	154	192	8	10	63	81	25	31	217	273
1996	11	23	137	171	3	8	61	72	14	31	198	243
1997	12	21	117	150	6	8	71	85	18	29	188	235
1998	19	30	125	174	6	11	68	85	25	41	193	259
1999	12	27	140	179	5	12	71	88	17	39	211	267
2000	22	29	133	184	7	17	67	91	29	46	200	275
2001	12	26	140	178	8	10	80	98	20	36	220	276
2002	10	23	156	189	5	13	85	103	15	36	241	292
2003	12	22	173	207	6	6	90	102	18	28	263	309

Fiscal Year	Application Shares (%)					Success Rates (%)			
	Woman	Minority	All others	All		Woman	Minority	All Others	All
1995	8.9	10.9	80.2	100.0		47.1	47.6	40.9	42.2
1996	6.4	13.5	80.1	100.0		27.3	34.8	44.5	42.1
1997	8.0	14.0	78.0	100.0		50.0	38.1	60.7	56.7
1998	10.9	17.2	71.8	100.0		31.6	36.7	54.4	48.9
1999	6.7	15.1	78.2	100.0		41.7	44.4	50.7	49.2
2000	12.0	15.8	72.3	100.0		31.8	58.6	50.4	49.5
2001	6.7	14.6	78.7	100.0		66.7	38.5	57.1	55.1
2002	5.3	12.2	82.5	100.0		50.0	56.5	54.5	54.5
2003	5.8	10.6	83.6	100.0		50.0	27.3	52.0	49.3

SOURCE: Department of Energy.

TABLE APP-A-25 Phase II Applications, by Company

Company	Number of Applications	Number of Awards	Phase II Dollars	Success Rate (%)
Top 20 Companies (Plus Ties)				
Physical Optics Corporation, Electro-optics & Holo	38	21	14,399,812	55.3
Eltron Research, Inc.	36	18	12,099,295	50.0
TDA Research, Inc.	34	21	14,850,271	61.8
Omega-p, Inc.	33	18	12,650,000	54.5
Radiation Monitoring Devices, Inc.	32	14	10,349,999	43.8
Physical Sciences, Inc.	32	12	7,966,660	37.5
Membrane Technology and Research, Inc.	26	14	9,923,405	53.8
Supercon, Inc.	24	8	4,599,731	33.3
MER Corporation	23	15	10,019,266	65.2
ADA Technologies, Inc.	23	9	6,299,009	39.1
Hypres, Inc.	22	10	6,599,415	45.5
Ceramem Corporation	21	13	9,200,000	61.9
Advanced Fuel Research, Inc.	20	8	5,136,686	40.0
Fm Technologies, Inc.	19	13	9,099,404	68.4
Spire Corporation	19	8	5,494,425	42.1
Science Research Laboratory, Inc.	18	12	7,949,428	66.7
Tech-x Corporation	18	11	7,843,709	61.1
American Superconductor Corporation	15	9	6,600,000	60.0
Calabazas Creek Research	15	12	7,395,865	80.0
Bend Research, Inc.	15	8	4,184,884	53.3
IGC Advanced Superconductors, Inc.	15	8	4,849,929	53.3
Other Top 20 awards winners (not 20 top applicants)				
Duly Research, Inc.	12	11	7,699,508	91.7
Diversified Technologies, Inc.	13	10	6,436,557	76.9
Haimson Research Corporation	9	9	6,079,745	100.0
Lynntech, Inc.	12	8	5,824,154	66.7
Fuelcell Energy, Inc.	11	8	5,249,873	72.7

SOURCE: Department of Energy.

Appendix B

NRC Phase II Survey

The first section of this appendix describes the methodology used to survey Phase II SBIR awards (also referred to as projects). The second part presents the results—of the awards, or project, survey (NRC Phase II Survey). (Appendix C presents the NRC Phase I survey.)

ABOUT THE SURVEYS

Starting Date and Coverage

The survey of SBIR Phase II awards was administered in 2005 and included awards made through 2001. This allowed most of the Phase II awarded projects (nominally two years) to be completed and provided some time for commercialization. The selection of the end date of 2001 was consistent with a GAO study, which in 1991 surveyed awards made through 1987.

A start date of 1992 was selected. The year 1992 for the earliest Phase II project was considered a realistic starting date for the coverage, allowing inclusion of the same (1992) projects as the Department of Defense (DoD) 1996 survey, and of the 1992 and 1993 projects surveyed in 1998 for the Small Business Administration (SBA). This adds to the longitudinal capacities of the study. The 10 years of Phase II coverage spanned the period of increased funding set-asides and the impact of the 1992 reauthorization. This time frame allowed for extended periods of commercialization and for a robust spectrum of economic conditions. Establishing 1992 as the cut-off date for starting the survey helped to avoid the problems from which older awards suffer, including meager early data collection as well as potentially irredeemable data loss; the fact that some firms and principal investigators (PIs) are no longer in place; and fading memories.

Award Numbers

While adding the annual awards numbers of the five agencies would seem to define the larger sample, the process was more complicated. Agency reports usually involve some estimating and anticipation of successful negotiation of selected proposals. Agencies rarely correct reports after the fact. Setting limitations on the number of projects to be surveyed from each firm required knowing how many awards each firm had received from all five agencies. Thus, the first step was to obtain all of the award databases from each agency and combine them into a single database. Defining the database was further complicated by variations in firm identification, location, phone numbers, and points of contact within individual agency databases. Ultimately, we determined that 4,085 firms had been awarded 11,214 Phase II awards (an average of 2.7 Phase II awards per firm) by the five agencies during the 1992-2001 time frame. Using the most recent awards, the firm information was updated to the most current contact information for each firm.

Sampling Approaches and Issues

The Phase II Survey used an array of sampling techniques to ensure adequate coverage of projects, to address a wide range both of outcomes and potential explanatory variables, and also to address the problem of skew. That is, a relatively small percentage of funded projects typically account for a large percentage of commercial impact in the field of advanced, high-risk technologies.

- **Random Samples.** After integrating the 11,214 awards into a single database, a random sample of approximately 20 percent was sampled. Then a random sample of 20 percent was ensured for each year; e.g., 20 percent of the 1992 awards, of the 1993 awards, etc. Verifying the total sample one year at a time allowed improved ability to adapt to changes in the program over time, as otherwise the increased number of awards made in recent years might dominate the sample.

- **Random Sample by Agency.** Surveyed awards were grouped by agency; additional respondents were randomly selected as required to ensure that at least 20 percent of each agency's awards were included in the sample.

- **Firm Surveys.** After the random selection, 100 percent of the Phase IIs that went to firms with only one or two awards were polled. These are the hardest firms to find for older awards. Address information is highly perishable, particularly for earlier award years. For firms that had more than two awards, 20 percent were selected, but no less than two.

- **Top Performers.** The problem of skew was dealt with by ensuring that all Phase IIs known to meet a specific commercialization threshold (total of $10 million in the sum of sales plus additional investment) were surveyed (derived from the DoD commercialization database). Since 56 percent of all awards were in the random and firm samples described above, only 95 Phase IIs were added in this fashion.

- **Coding.** The project database tracks the survey sample, which corresponds with each response. For example, it is possible for a randomly sampled project from a firm that had only two awards to be a top performer. Thus, the response could be analyzed as a random sample for the program, a random sample for the awarding agency, a top performer, and as part of the sample of single or double winners. In addition, the database allows examination of the responses for the array of potential explanatory or demographic variables.

- **Total Number of Surveys.** The approach described above generated a sample of 6,410 projects and 4,085 firm surveys—an average of 1.6 award surveys per firm. Each firm receiving at least one project survey also received a firm survey.[1] Although this approach sampled more than 57 percent of the awards, multiple award winners, on average, were asked to respond to surveys covering about 20 percent of their projects.

Administration of the Survey

The questionnaire drew extensively from the one used in the 1999 National Research Council assessment of SBIR at the Department of Defense, *The Small Business Innovation Research Program: An Assessment of the Department of Defense Fast Track Initiative.*[2] That questionnaire in turn built upon the questionnaire for the 1991 GAO SBIR study. Twenty-four of the 29 questions on the earlier NRC study were incorporated. The researchers added 24 new questions to attempt to understand both commercial and noncommercial aspects, including knowledge base impacts, of SBIR, and to gain insight into impacts of program management. Potential questions were discussed with each agency, and their

---

[1] For NRC Firm Survey results, see National Research Council, *An Assessment of The Small Business Innovation Research Program*, Charles W. Wessner, ed., Washington, DC: The National Academies Press, 2008.

[2] National Research Council, *The Small Business Innovation Research Program: An Assessment of the Department of Defense Fast Track Initiative*, Charles W. Wessner, ed., Washington, DC: National Academy Press, 2000.

input was considered. In determining questions that should be in the survey, the research team also considered which issues and questions were best examined in the case studies and other research methodologies. Many of the resultant 32 Phase II Survey questions and 15 Firm Survey questions had multiple parts.

The surveys were administered online, using a Web server. The formatting, encoding and administration of the survey was subcontracted to BRTRC, Inc., of Fairfax, Virginia.

There are many advantages to online surveys (including cost, speed, and possibly response rates). Response rates become clear fairly quickly, and can rapidly indicate needed follow up for nonrespondents. Hyperlinks provide amplifying information, and built-in quality checks control the internal consistency of the responses. Finally, online surveys allow dynamic branching of question sets, with some respondents answering selected subsets of questions but not others, depending on prior responses.

Prior to the survey, we recognized two significant advantages of a paper survey over an online one. For every firm (and thus every award), the agencies had provided a mailing address. Thus surveys could be addressed to the firm president or CEO at that address. That senior official could then forward the survey to the correct official within the firm for completion. For an online survey we needed to know the email address of the correct official. Also each firm needed a password to protect its answers. We had an SBIR point of contact (POC) and an email address and a password for every firm, which had submitted for a DoD SBIR 1999 survey. However, we had only limited email addresses and no passwords for the remainder of the firms. For many, the email addresses that we did have were those of principal investigators rather than an official of the firm. The decision to use an online survey meant that the first step of survey distribution was an outreach effort to establish contact with the firms.

Outreach by Mail

This outreach phase began with establishing a National Academy of Sciences (NAS) registration Web site which allowed each firm to establish a POC, an email address, and a password. Next, the Study Director, Dr. Charles Wessner, sent a letter to those firms for which email contacts were not available. Ultimately, only 150 of the 2,080[3] firms provided POC/email after receipt of this letter. The U.S. Postal Service returned 650 of those letters as invalid addresses. Each returned letter required thorough research by calling the agency-provided phone number for the firm, then using the Central Contractor Registration database, Business.com (powered by Google), and Switchboard.com to try to find correct address information. When an apparent match was found, the firm was called to verify that it was, in fact, the firm, which had completed the SBIR. Two hundred

---

[3]The letter was also erroneously sent to an additional 43 firms that had received only STTR awards.

thirty-seven of the 650 missing firms were so located. Another 10 firms that had gone out of business and had no POC were located.

Two months after the first mailing, a second letter from the study director was mailed to firms whose first letter had not been returned, but which had not yet registered a POC. This letter also went to 176 firms, which had a POC email but no password, and to the 237 newly corrected addresses. The large number of letters (277) from this second mailing that were returned by the U.S. Postal Service indicated that there were more bad addresses in the first mailing than were indicated by its returned mail. (If the initial letter was inadvertently delivered, it may have been thrown away.) Of the 277 returned second letters, 58 firms were located using the search methodology described above. These firms were asked on the phone to go to the registration Web site to enter POC/email/password. A total of 93 firms provided POC/email/password on the registration site subsequent to the second mailing. Three additional firms were identified as out of business.

The final mailing, a week before survey, was sent to those firms that had not received either of the first two letters. It announced the study/survey and requested support of the 1,888 CEOs for which we had assumed good POC/email information from the DoD SBIR submission site. That letter asked the recipients to provide new contact information at the DoD submission site if the firm information had changed since their last submission. One hundred seventy-three of these letters were returned. We were able to find new addresses for 53 of these, and to ask those firms to update their information. One hundred fifteen firms could not be found, and 5 more were identified as out of business.

The three mailings had demonstrated that at least 1,100 (27 percent) of the mailing addresses were in error, 734 of which firms could not be found, and 18 were reported to be out of business.

Outreach by Email

We began Internet contact by emailing the 1,888 DoD POCs to verify their email addresses and to give them an opportunity to identify a new POC. Four hundred ninety-four of those emails bounced. The next email went to 788 email addresses that we had received from agencies as PI email addresses. We asked that the PI have the correct company POCs identify themselves at the NAS Update registration site. One hundred eighty-eight of these emails bounced. After a more detailed search of the list used by the National Institutes of Health (NIH) to send out their survey, we identified 83 additional PIs and sent them the PI email discussed above. Email to the POCs not on the DoD Submission site resulted in 110 more POCs/emails/passwords being registered on the NAS registration site.

We began the survey at the end of February with an email to 100 POCs as a beta test and followed that with another email to 2,041 POCs (total of 2,141) a week later.

Survey Responses

By August 5, 2005, five months after release of the survey, 1,239 firms had begun and 1,149 firms had completed at least 14 of 15 questions on the Firm Survey. Project surveys were begun on 1,916 Phase II awards. Of the 4,085 firms that received Phase II SBIR awards from DoD, NIH, NASA, NSF, or DoE from 1992 to 2001, an additional 7 firms were identified as out of business (total of 25), and no email addresses could be found for 893. For an additional 500 firms, the best email addresses that were found were also undeliverable. These 1,418 firms could not be contacted and thus had no opportunity to complete the surveys. Of these firms, 585 had mailing addresses known to be bad. The 1,418 firms that could not be contacted were responsible for 1,885 of the individual awards in the sample.

Using the same methodology as the GAO had used in the 1992 report of their 1991 survey of SBIR, undeliverables and out of business firms were eliminated prior to determining the response rate. Although 4,085 firms were surveyed, 1,418 firms were eliminated as described. This left 2,667 firms, of which 1,239 responded, representing a 46 percent response rate by firms,[4] which could respond. Similarly, when the awards, which were won by firms in the undeliverable category, were eliminated (6,408 minus 1,885), this left 4,523 projects, of which 1,916 responded, representing a 42 percent response rate. Table App-B-1 displays by agency the number of Phase II awards in the sample, the number of those awards, which by having good email addresses had the opportunity to respond, and the number that responded.[5] Percentages displayed are the percentage of awards with good addresses, the percentage of the sample that responded, and the responses as a percentage of awards with the opportunity to respond.

The NRC Methodology report had assumed a response rate of about 20 percent. Considering the length of the survey and its voluntary nature, the rate achieved was relatively high and reflects both the interest of the participants in the SBIR program and the extensive follow-up efforts. At the same time, the possibility of response biases that could significantly affect the survey results must be recognized. For example, it may be possible that some of the firms that could not be found have been unsuccessful and folded. It may also be possible that unsuccessful firms were less likely to respond to the survey.

---

[4]Firm information and response percentages are not displayed in Table App-B-1, which displays by agency, since many firms received awards from multiple agencies.

[5]The average firm size for awards, which responded, was 37 employees. Nonresponding awards came from firms that averaged 38 employees. Since responding Phase IIs were more generally more recent than nonresponding, and awards have gradually grown in size, the difference in average award size ($655,525 for responding and $649,715 for nonresponding) seems minor.

TABLE App-B-1 NRC Phase II Survey Responses by Agency as of August 4, 2005

Agency	Phase II Sample Size	Awards with Good Email Addresses	Percentage of Sample Awards with Good Email Addresses	Answered Survey as of 8/4/2005	Surveys as a Percentage of Sample	Surveys as a Percentage of Awards Contacted
DoD	3,055	2,191	72	920	30	42
NIH	1,680	1,127	67	496	30	44
NASA	779	534	69	181	23	34
NSF	457	336	74	162	35	48
DoE	439	335	76	157	36	47
Total	6,408	4,523	70	1,916	30	42

NRC Phase II Survey Results

NOTE: RESULTS ARE FOR DOE. SURVEY RESPONSES APPEAR IN BOLD, AND EXPLANATORY NOTES ARE IN A TYPEWRITER FONT.

Project Information 157 respondents answered the first question. Since respondents are directed to skip certain questions based on prior answers, the number that responded varies by question. Also, some respondents did not complete their surveys. 148 completed all applicable questions. For computation of averages, such as average sales, the denominator used was 157, the number of respondents who answered the first question. Where appropriate, the basis for calculations is provided after the question.

PROPOSAL TITLE:
AGENCY: DoE
TOPIC NUMBER:
PHASE II CONTRACT/GRANT NUMBER:

Part I. Current status of the Project.

1. What is the current status of the project funded by the referenced SBIR award? *Select the one best answer.* Percentages are based on the 157 respondents who answered this question.
   a. **3%** Project has not yet completed Phase II. *Go to question 21.*
   b. **21%** Efforts at this company have been discontinued. No sales or additional funding resulted from this project. *Go to question 2.*
   c. **17%** Efforts at this company have been discontinued. The project did result in sales, licensing of technology, or additional funding. *Go to question 2.*
   d. **17%** Project is continuing post-Phase II technology development. *Go to question 3.*
   e. **18%** Commercialization is underway. *Go to question 3.*
   f. **24%** Products/Processes/ Services are in use by target population/customer/consumers. *Go to question 3.*

2. Did the reasons for discontinuing this project include any of the following? *PLEASE SELECT YES OR NO FOR EACH REASON AND NOTE THE ONE PRIMARY REASON.*
59 projects were discontinued. The % below is the percent of the discontinued projects that responded with the indicated response.

APPENDIX B

	Yes	No	Primary Reason
a. Technical failure or difficulties	41%	59%	25%
b. Market demand too small	54%	46%	32%
c. Level of technical risk too high	24%	76%	2%
d. Not enough funding	53%	47%	12%
e. Company shifted priorities	27%	73%	7%
f. Principal investigator left	10%	90%	3%
g. Project goal was achieved (e.g., prototype delivered for federal agency use)	49%	51%	5%
h. Licensed to another company	10%	90%	7%
i. Product, process, or service not competitive	31%	69%	5%
j. Inadequate sales capability	10%	90%	0%
k. Other (please specify): _____	7%	93%	2%

The next question to be answered depends on the answer to Question 1. If c., go to Question 3. If b, skip to Question 16.

Part II. Commercialization activities and planning.

Questions 3-7 concern actual sales to date resulting from the technology developed during this project. **Sales** includes all sales of a product, process, or service, to federal or private sector customers resulting from the technology developed during this Phase II project. A sale also includes licensing, the sale of technology or rights, etc.

3. Has your company and/or licensee had any actual sales of products, processes, services or other sales incorporating the technology developed during this project? *Select all that apply.* This question was not answered for those projects still in Phase II (3%) or for projects, which were discontinued without sales or additional funding (21%). The denominator for the percentages below is all projects that answered the survey. Only 76% of all projects, which answered the survey, could respond to this question.

   a. **16%** No sales to date, but sales are expected. *Skip to Question 8*
   b. **7%** No sales to date nor are sales expected. *Skip to Question 11*
   c. **39%** Sales of product(s)
   d. **9%** Sales of process(es)
   e. **22%** Sales of services(s)
   f. **11%** Other sales (e.g. rights to technology, licensing, etc.)

   From the combination of responses 1b, 3a and 3b, we can conclude that 28% had no sales and expect none, and that 18% had no sales but expect sales.

4. For your company and/or your licensee(s), when did the first sale occur, and what is the approximate amount of total sales resulting from the technology developed during this project? If multiple SBIR awards contributed to the ultimate commercial outcome, report only the share of total sales appropriate to this SBIR project. *Enter the requested information for your company in the first column and, if applicable and if known, for your licensee(s) in the second column. Enter approximate dollars. If none, enter 0 (zero)*

	Your Company	Licensee(s)
a. Year when first sale occurred	☐☐☐☐	☐☐☐☐

   50% reported a year of first sale. 50% of these first sales occurred in 2000 or later. 19% reported a licensee year of first sale. 67% of these first sales occurred in 2000 or later.

   b. Total Sales Dollars of Product (s) Process(es) **$ 582,783**   **$267,535**
      or Service(s) to date. (average of 157 survey respondents)

   Although 82 reported a year of first sale, only 74 reported sales >0. Their average sales were $ 1,115,817. 47% of the total sales dollars were due to 4 projects, each of which had $5,500,000 or more in sales. The highest reporting project had $17,515,000 in sales. Similarly of the 30 projects that reported a year of first licensee sale, only 7 reported actual licensee sales >0. Their average sales were $6,000,429. 76% of the total sales dollars was due to projects, each of which had 23,000,000 or more licensee sales. The highest reporting project had 31,926,000 in licensee sales.

   c. Other Total Sales Dollars (e.g., Rights to   **$ 45,556**   **$ 2,675**
      technology, Sale of spin-off company, etc.) to date (average of 157 survey respondents)

   Combining the responses for b and c, the average for each of the 157 projects that responded to the survey is thus sales of over $625,000 by the SBIR company and over $270,000 in sales by licensees.

Display this box for Questions 4 & 5 if project commercialization is known.
Your company reported sales information to DoD as a part of an SBIR proposal or to NRC as a result of an earlier NRC request. This information may be useful in answering the prior question or the next question. You reported as of *(date)*: DoD sales *($ amount),* Other federal sales *($ amount),* Export sales *($ amount),* Private-Sector sales *($ amount)*, and other sales *($ amount).*

APPENDIX B

5. To date, approximately what percent of total sales from the technology developed during this project have gone to the following customers? *If none enter 0 (zero). Round percentages. Answers should add to about 100%.*[6] 157 firms responded to this question as to what percent of their sales went to each agency or sector.

Domestic private sector	**76%**
Department of Defense (DoD)	**2%**
Prime contractors for *DoD or NASA*	**1%**
NASA	**1%**
Agency that awarded the Phase II	**–%**
Other federal agencies *(Pull down)*	**1%**
State or local governments	**0%**
Export markets	**14%**
Other *(Specify)*_____	**1%**

The following questions identify the product, process, or service resulting from the project supported by the referenced SBIR award, including its use in a fielded federal system or a federal acquisition program.

---

6. Is a federal system or acquisition program using the technology from this Phase II? If yes, please provide the name of the federal system or acquisition program that is using the technology. **5% reported use in a federal system or acquisition program.**

7. Did a commercial product result from this Phase II project? **15% reported a commercial product.**

8. If you have had no sales to date resulting from the technology developed during this project, what year do you expect the first sales for your company or its licensee? Only firms that had no sales but answered that they expected sales got this question.

    29% expected sales. The year of expected first sale is ☐☐☐☐
    71% of those expecting sales expected sales to occur before 2007.

---

[6]Please note: If a NASA SBIR award, the prime contractor's line will state "Prime contractors for NASA." The "Agency that awarded the Phase II" will only appear if it is not DoD or NASA. The name of the actual awarding agency will appear.

9. For your company and/or your licensee, what is the approximate amount of total sales expected between now and the end of 2006 resulting from the technology developed during this project? *If none, enter 0 (zero).* This question was seen by those who already had sales and those w/o sales who reported expecting sales; however, averages are computed for all who took the survey since all could have expected sales.

   a. Total sales dollars of product(s), process(es) or services(s) expected between now and the end of 2006. **$ 551,949** (average of 157 projects)

   b. Other Total Sales Dollars (e.g., rights to technology, sale of spin-off company, etc.) expected between now and the end of 2006. **$ 107,325** (average of 157 projects)

   c. Basis of expected sales estimate. *(Select all that apply.)*
      - a. **12%** Market research
      - b. **25%** Ongoing negotiations
      - c. **44%** Projection from current sales
      - d. **2%** Consultant estimate
      - e. **30%** Past experience
      - f. **33%** Educated guess

10. How did you (or do you expect to) commercialize your SBIR award?
    - a. **5%** No commercial product, process, or service was/is planned.
    - b. **19%** As software
    - c. **58%** As hardware (final product, component, or intermediate hardware product)
    - d. **32%** As process technology
    - e. **21%** As new or improved service capability
    - f. **0%** As a drug
    - g. **0%** As a biologic
    - h. **14%** As a research tool
    - i. **4%** As educational materials
    - j. **8%** Other, please explain _____

11. Which of the following, if any, describes the type and status of marketing activities by your company and/or your licensee for this project? *Select one for each marketing activity.* This question answered by 117 firms, which completed Phase II and have not discontinued the project, w/o sales or additional funding.

Marketing activity	Planned	Need Assistance	Underway	Completed	Not Needed
a. Preparation of marketing plan	4%	3%	15%	44%	34%
b. Hiring of marketing staff	3%	5%	9%	22%	61%
c. Publicity/advertising	9%	6%	26%	29%	30%
d. Test marketing	8%	5%	20%	18%	50%
e. Market research	5%	5%	22%	34%	33%
f. Other (Specify)	0%	0%	3%	3%	39%

Part III. Other outcomes.

12. As a result of the technology developed during this project, which of the following describes your company's activities with other companies and investors? *Select all that apply.* Percentage of the 117 who answered this question.

Activities	U.S. Companies/Investors		Foreign Companies/Investors	
	Finalized Agreements	Ongoing Negotiations	Finalized Agreements	Ongoing Negotiations
a. Licensing Agreement(s)	15%	15%	5%	9%
b. Sale of company	1%	1%	1%	1%
c. Partial sale of company	3%	2%	0%	2%
d. Sale of technology rights	5%	6%	1%	3%
e. Company merger	0%	1%	0%	1%
f. Joint Venture agreement	3%	7%	0%	3%
g. Marketing/distribution agreement(s)	9%	7%	9%	3%
h. Manufacturing agreement(s)	6%	6%	1%	4%
i. R&D agreement(s)	8%	10%	2%	6%
j. Customer alliance(s)	8%	13%	4%	5%
k. Other (Specify) _____	5%	3%	0%	1%

13. In your opinion, in the absence of this SBIR award, would your company have undertaken this project?
    Select one. Percentage of the 117 who answered this question.
    a. **1%** Definitely yes
    b. **4%** Probably yes *(If selected a. or b., Go to question 14.)*
    c. **13%** Uncertain
    d. **44%** Probably not
    e. **38%** Definitely not *(If c., d. or e., skip to question 16.)*

14. If you had undertaken this project in the absence of SBIR, this project would have been Questions 14 and 15 were answered only by the 5% who responded that they definitely or probably would have undertaken this project in the absence of SBIR.
    a. **0%** Broader in scope
    b. **17%** Similar in scope
    c. **83%** Narrower in scope

15. In the absence of SBIR funding, (Please provide your best estimate of the impact.)
    a. The start of this project would have been delayed about **an average of 11** months.
    50% of the 6 firms expected the project would have been delayed. 33% anticipated a delay of at least 24 months.
    b. The expected duration/time to completion would have been
        1) **83%** Longer
        2) **0%** The same
        3) **0%** Shorter
        17% No response
    c. In achieving <u>similar</u> goals and milestones, the project would be
        1) **0%** Ahead
        2) **0%** The same place
        3) **83%** Behind
        17% No response

16. Employee information. *Enter number of employees. You may enter fractions of full-time effort (e.g., 1.2 employees). Please include both part-time and full-time employees, and consultants, in your calculation.*

Number of employees (if known) when Phase II proposal was submitted.	**Ave = 32** **1% report 0** **30% report 1 - 5** **28% report 6 - 20** **18% report 21 - 50** **10% report >100**
Current number of employees.	**Ave = 54** **2% report 0** **17% report 1 - 5** **33% report 6 - 20** **22% report 21 - 50** **14% report >100**
Number of current employees <u>who were hired</u> as a result of the technology developed during this Phase II project.	**Ave = 1.5** **46% report 0** **49% report 1 - 5** **5% report 6 – 20** **0% report >20**
Number of current employees <u>who were retained</u> as a result of the technology developed during this Phase II project.	**Ave = 1.5** **43% report 0** **52% report 1 - 5** **3% report 6 – 20** **0% report >20**

17. The Principal Investigator for this Phase II award was a (check all that apply)
    a. **5%** Woman
    b. **11%** Minority
    c. **85%** Neither a woman or minority

18. Please give the number of patents, copyrights, trademarks, and/or scientific publications for the technology developed as a result of this project. *Enter numbers. If none, enter 0 (zero).* Results are for 149 respondents to this question.

Number Applied For/Submitted		Number Received/Published
123	Patents	91
21	Copyrights	20
31	Trademarks	27
218	Scientific Publications	200

Part IV. Other SBIR funding.

19. How many SBIR awards did your company receive prior to the Phase I that led to this Phase II?
    a. Number of previous Phase I awards. **Average of 18. 28% had no prior Phase I, and another 36% had 5 or less prior Phase I.**
    b. Number of previous Phase II awards. **Average of 7. 40% had no prior Phase II, and another 36% had 5 or less prior Phase II.**

20. How many SBIR awards has your company received that are related to the project/technology supported by this Phase II award?
    a. Number of related Phase I awards   **Average of 2; 50% had no prior related Phase I and another 42% had 5 or less prior related Phase I**
    b. Number of related Phase II awards   **Average of 1; 62% had no prior related Phase II and another 37% had 5 or less prior related Phase II**

Part V. Funding and other assistance.

21. Prior to this SBIR Phase II award, did your company receive funds for research or development of the technology in this project from any of the following sources? Of 152 respondents.
    a. **26%** Prior SBIR *(Excluding the Phase I, which proceeded this Phase II.)*
    b. **13%** Prior non-SBIR federal R&D
    c. **3%** Venture Capital
    d. **8%** Other private company
    e. **5%** Private investor
    f. **30%** Internal company investment (including borrowed money)
    g. **3%** State or local government
    h. **3%** College or University
    i. **3%** Other *Specify* _____

Commercialization of the results of an SBIR project normally requires additional developmental funding. Questions 22 and 23 address additional funding. Additional Developmental Funds include non-SBIR funds from federal or private sector sources, or from your own company, used for further development and/or commercialization of the technology developed during this Phase II project.

---

22. Have you received or invested any additional developmental funding in this project?

    a. **63%** Yes *Continue.*
    b. **37%** No *Skip to Question 24.*

23. To date, what has been the total additional developmental funding for the technology developed during this project? Any entries in the **Reported** column are based on information previously reported by your firm to DoD or NAS. They are provided to assist you in completing the **Developmental funding** column. Previously reported information did not include investment by your company or personal investment. *Please update this information to include breaking out Private investment and Other investment by subcategory. Enter dollars provided by each of the listed sources. If none, enter 0 (zero).* The dollars shown are determined by dividing the total funding in that category by the 157 respondents who started the survey to determine an average funding. Ninety-three of these respondents reported any additional funding.

Source	Reported	Developmental Funding
a. Non-SBIR federal funds	$_ _, _ _ _, _ _ _	**$ 362,968**
b. Private investment	$_ _, _ _ _, _ _ _	
(1) U.S. venture capital		**$ 0**
(2) Foreign investment		**$ 124,522**
(3) Other private equity		**$ 36,908**
(4) Other domestic private company		**$ 224,358**
c. Other sources	$_ _, _ _ _, _ _ _	
(1) State or local governments		**$ 2,025**
(2) College or universities		**$ 817**
d. Not previously reported		
(1) Your own company (Including money you have borrowed)		**$ 156,621**
(2) Personal funds		**$ 13,216**
Total average additional developmental funding, all sources, per award		$ 921,438

24. Did this award identify matching funds or other types of cost sharing in the Phase II Proposal?[7]
    a. **58%** No matching funds/co-investment/cost sharing were identified in the proposal.
       If a, skip to question 26.
    b. **42%** <u>Although not a DoD Fast Track</u>, matching funds/co-investment/cost sharing were identified in the proposal.
    c. **0%** <u>Yes. This was a DoD Fast Track proposal.</u>

25. Regarding sources of matching or co-investment funding that were proposed for Phase II, check all that apply. The percentages below are computed for those 63 projects, which reported matching funds.
    a. **73%** Our own company provided funding (includes borrowed funds).
    b. **13%** A federal agency provided non-SBIR funds.
    c. **43%** Another company provided funding.
    d. **2%** An angel or other private investment source provided funding.
    e. **5%** Venture Capital provided funding.

26. Did you experience a gap between the end of Phase I and the start of Phase II?
    a. **64%** Yes *Continue.*
    b. **36%** No *Skip to question 29.*
    The average gap reported by 95 respondents was 5 months. 3% of the respondents reported a gap of one or more years.

27. Project history. Please fill in for all dates that have occurred. This information is meaningless in aggregate. It has to be examined project by project in conjunction with the date of the Phase I end and the date of the Phase II award to calculate the gaps.

    Date Phase I ended *Month/ year*   ☐☐☐☐

    Date Phase II proposal submitted *Month /year*   ☐☐☐☐

28. If you experienced funding gap between Phase I and Phase II for this award, *select all answers that apply.*
    a. **56%** Stopped work on this project during funding gap.
    b. **33%** Continued work at reduced pace during funding gap.
    c. **9%** Continued work at pace equal to or greater than Phase I pace during funding gap.
    d. **3%** Received bridge funding between Phase I and II.
    e. **3%** Company ceased all operations during funding gap.

---

[7]The words <u>underlined</u> appear only for DoD awards.

APPENDIX B

29. Did you receive assistance in Phase I or Phase II proposal preparation for this award? Of 148 respondents.
    a. **2%** State agency provided assistance.
    b. **3%** Mentor company provided assistance.
    c. **1%** Regional association provided assistance.
    d. **8%** University provided assistance.
    e. **86%** We received no assistance in proposal preparation.

    Was this assistance useful?
    a. **60%** Very useful
    b. **35%** Somewhat useful
    c. **5%** Not useful

30. In executing this award, was there any involvement by university faculty, graduate students, and/or university developed technologies? Of 148 respondents.
    38% Yes
    63% No

31. This question addresses any relationships between your firm's efforts on this Phase II project and any University (ies) or College (s). The percentages are computed against the 148 who answered question 30, not just those who answered yes to question 30. *Select all that apply.*
    a. **0%** The Principal Investigator (PI) for this Phase II project was at the time of the project a faculty member.
    b. **3%** The Principal Investigator (PI) for this Phase II project was at the time of the project an adjunct faculty member.
    c. **23%** Faculty member(s) or adjunct faculty member (s) work on this Phase II project in a role other than PI, e.g., consultant.
    d. **20%** Graduate students worked on this Phase II project.
    e. **16%** University/College facilities and/or equipment were used on this Phase II project.
    f. **1%** The technology for this project was licensed from a University or College.
    g. **5%** The technology for this project was originally developed at a University or College by one of the percipients in this phase II project.
    h. **18%** A University or College was a subcontractor on this Phase II project.

    In remarks enter the name of the University or College that is referred to in any blocks that are checked above. If more than one institution is referred to, briefly indicate the name and role of each.

32. Did commercialization of the results of your SBIR award require FDA approval? Yes **1%**

    In what stage of the approval process are you for commercializing this SBIR award?
    a. **0%** Applied for approval
    b. **0%** Review ongoing
    c. **0%** Approved
    d. **0%** Not approved
    e. **0%** IND: Clinical trials
    f. **0%** Other

Appendix C

NRC Phase I Survey

SURVEY DESCRIPTION

This section describes a survey of Phase I SBIR awards over the period 1992–2001. The intent of the survey was to obtain information on those which did not proceed to Phase II, although most that did receive a Phase II were also surveyed.

Over that period the five agencies (DoD, DoE, NIH, NASA, and NSF) made 27,978 Phase I awards. Of the total number for the five agencies, 7,940 Phase I awards could be linked to one of the 11,214 Phase II awards made from 1992–2001. To avoid putting an unreasonable burden on the firms which had many awards, we identified all firms which had over 10 Phase I awards that apparently had not received a Phase II. For those firms, we did not survey any Phase I awards that also received a Phase II. This amounted to 1,679 Phase Is that were not surveyed.

We chose to survey the principal investigator (PI) rather than the firm both to reduce the number of surveys that any person would have to complete, and because if the Phase I had not gone on to a Phase II, the PI was more likely to have any memory of it than would the firm officials. There were no PI email addresses for 5,030 Phase Is, a fact that reduced the number of surveys sent since the survey was conduced by email.

Thus there were 21,269 surveys (27,978 minus 1,679 minus 5,030 = 21,269) emailed to 9,184 PIs. Many PIs had received multiple Phase Is. Of these surveys, 6,770 were bounced (undeliverable) email. This left possible responses of 14,499. Of these, there were 2,746 responses received. The responses received represented 9.8 percent of all Phase I awards for the five agencies, or 12.9 percent of all surveys emailed, and 18.9 percent of all possible responses.

The agency breakdown, including Phase I Survey results, is given in Table App-C-1.

TABLE App-C-1 Agency Breakdown for Phase I Survey

Phase I Surveys by Agency	Phase I awards, 1992–2001	Answered Survey (Number)	Answered Survey (%)
DoD	13,103	1,198	9
DoE	2,005	281	14
NASA	3,363	303	9
NIH	7,049	716	10
NSF	2,458	248	10
Total	27,978	2,746	10

APPENDIX C

SURVEY PREFACE

This survey is an important part of a major study commissioned by the U.S. Congress to review the SBIR program as it is operated at various federal agencies. The assessment, by the National Research Council (NRC), seeks to determine both the extent to which the SBIR programs meet their mandated objectives, and to investigate ways in which the programs could be improved. Over 1,200 firms have participated earlier this year in extensive survey efforts related to firm dynamics and Phase II awards. This survey attempts to determine the impact of Phase I awards that do not go on to Phase II. We need your help in this assessment. We believe that you were the PI on the listed Phase I.

We anticipate that the survey will take about 5-10 minutes of your time. If this Phase I resulted in a Phase II, this survey has only three questions; if there was not a Phase II; there are 14 questions. Where $ figures are requested (sales or funding,) please give your best estimate. Responses will be aggregated for statistical analysis and not attributed to the responding firm/PI, without the subsequent explicit permission of the firm.

Since you have been the PI on more than one Phase I from 1992 to 2001, you will receive additional surveys. These are not duplicates. Please complete as many surveys for those Phase I that did not result in a Phase II as you deem to be reasonable.

Further information on the study can be found at *<http://www7.nationalacademies.org/sbir>*. BRTRC, Inc is administering this survey for the NRC. If you need assistance in completing the survey, call 877-270-5392. If you have questions about the assessment more broadly, please contact Dr. Charles Wessner, Study Director, NRC.

Project Information
Proposal Title:
Agency:
Firm Name:
Phase I Contract/Grant Number:

NRC Phase I Survey Results

NOTE: ALL RESULTS APPEAR IN BOLD. RESULTS ARE REPORTED FOR ALL 5 AGENCIES (DOD, NIH, NSF, DOE, AND NASA). EXPLANATORY NOTES ARE IN A TYPEWRITER FONT.

2,746 responded to the survey. Of these 1,380 received the follow on Phase II. 1,366 received only a Phase I.

1. Did you receive assistance in preparation for this Phase I proposal?

Phase I only		Received Phase II	
95%	No *Skip to Question 3.*	**93%**	No
5%	Yes *Go to Question 2.*	**7%**	Yes

2. If you received assistance in preparation for this Phase I proposal, put an X in the first column for any sources that assisted and in the second column for the most useful source of assistance. Check all that apply. Answered by 74 Phase I only and 91 Phase II who received assistance.

Phase I only		Received Phase II
Assisted/Most Useful		Assisted/Most Useful
10/3	State agency provided assistance	**11/10**
15/9	Mentor company provided assistance	**21/15**
31/17	University provided assistance	**34/22**
16/8	Federal agency SBIR program managers or technical representatives provided assistance	**25/19**

3. Did you receive a Phase II award as a sequential direct follow-on to this Phase I award? *If yes, please check yes. Your survey would have been automatically submitted with the HTML format. Using this Word format, you are done after answering this question. Please email this as an attachment to jcahill@brtrc.com or fax to Joe Cahill 703 204 9447. Thank you for you participation.* 2,746 responses.

   50% No. We did not receive a follow-on Phase II after this Phase I.
   50% Yes. We did receive the follow-on Phase II after this Phase I.

APPENDIX C

4. Which statement correctly describes why you did not receive the Phase II award after completion of your Phase I effort. *Select best answer.* All questions which follow were answered by those 1,366 who did not receive the follow on Phase II. % based on 1,366 responses.

33%	The company did not apply for a Phase II. Go to question 5.
63%	The company applied, but was not selected for a Phase II. Skip to question 6.
1%	The company was selected for a Phase II, but negotiations with the government failed to result in a grant or contract. Skip to question 6.
3%	Did not respond to question 4.

5. The company did not apply for a Phase II because: *Select all that apply.* % based on 446 who answered "The company did not apply for a Phase II" in question 4.

38%	Phase I did not demonstrate sufficient technical promise.
11%	Phase II was not expected to have sufficient commercial promise.
6%	The research goals were met by Phase I. No Phase II was required.
34%	The agency did not invite a Phase II proposal.
3%	Preparation of a Phase II proposal was considered too difficult to be cost effective.
1%	The company did not want to undergo the audit process.
8%	The company shifted priorities.
5%	The PI was no longer available.
6%	The government indicated it was not interested in a Phase II.
13%	Other—explain:

6. Did this Phase I produce a noncommercial benefit? Check all responses that apply. % based on 1,366.

59%	The awarding agency obtained useful information.
83%	The firm improved its knowledge of this technology.
27%	The firm hired or retained one or more valuable employees.
17%	The public directly benefited or will benefit from the results of this Phase I. *Briefly explain benefit.*
13%	This Phase I was essential to founding the firm or to keeping the firm in business.
8%	No

7. Although no Phase II was awarded, did your company continue to pursue the technology examined in this Phase I? *Select all that apply.* % based on 1,366.

46%	The company did not pursue this effort further.
22%	The company received at least one subsequent Phase I SBIR award in this technology.
14%	Although the company did not receive the direct follow-on Phase II to this Phase I, the company did receive at least one other subsequent Phase II SBIR award in this technology.
12%	The company received subsequent federal non-SBIR contracts or grants in this technology.
9%	The company commercialized the technology from this Phase I.
2%	The company licensed or sold their rights in the technology developed in this Phase I.
16%	The company pursued the technology after Phase I, but it did not result in subsequent grants, contracts, licensing or sales.

Part II. Commercialization

8. How did you, or do you, expect to commercialize your SBIR award? *Select all that apply.* % based on 1,366.

33%	No commercial product, process, or service was/is planned.
16%	As software
32%	As hardware (final product component or intermediate hardware product)
20%	As process technology
11%	As new or improved service capability
15%	As a research tool
4%	As a drug or biologic
3%	As educational materials

9. Has your company had any actual sales of products, processes, services, or other sales incorporating the technology developed during this Phase I? *Select all that apply.* % based on 1,366.

5%	Although there are no sales to date, the outcome of this Phase I is in use by the intended target population.
65%	No sales to date, nor are sales expected. Go to question 11.
15%	No sales to date, but sales are expected. Go to question 11.
9%	Sales of product(s)

1%	Sales of process(es)
6%	Sales of services(s)
2%	Other sales (e.g., rights to technology, sale of spin-off company, etc.)
2%	Licensing fees

10. For your company and/or your licensee(s), when did the first sale occur, and what is the approximate amount of total sales resulting from the technology developed during this phase I? If other SBIR awards contributed to the ultimate commercial outcome, estimate only the share of total sales appropriate to this Phase I project. (Enter the requested information for your company in the first column and, if applicable and if known, for your licensee(s) in the second column. Enter dollars. If none, enter 0 [zero]; leave blank if unknown.)

	Your Company	Licensee(s)
a. Year when first sale occurred	89 of 147 after 1999	11 of 13 after 1999
b. Total Sales Dollars of Product(s) Process(es) or Service(s) to date		
Sale Averages	**$84,735**	**$3,947**
Top 5 Sales Accounts for **43%** of all sales	1. **$20,000,000** 2. **$15,000,000** 3. **$5,600,000** 4. **$5,000,000** 5. **$4,200,000**	
c. Other Total Sales Dollars (e.g., Rights to technology, Sale of spin-off company, etc.) to date		
Sale Averages	**$1,878**	**$0**

Sale averages determined by dividing totals by 1,366 responders

11. If applicable, please give the number of patents, copyrights, trademarks and/or scientific publications for the technology developed as a result of Phase I. (Enter numbers. If none, enter 0 [zero]; leave blank if unknown.)

#Applied For or Submitted / # Received/Published

319 / 251 Patent(s)
50 / 42   Copyright(s)
52 / 47   Trademark(s)
521 / 472 Scientific Publication(s)

12. In your opinion, in the absence of this Phase I award, would your company have undertaken this Phase I research? *Select only one lettered response. If you select c., and the research, absent the SBIR award, would have been different in scope or duration, check all appropriate boxes.* Unless otherwise stated, % are based on 1,366.

    5%   Definitely yes
    7%   Probably yes, similar scope and duration
    16% Probably yes, but the research would have been different in the following way
              % based on 218 who responded probably yes, but research would have . . .
              75% Reduced scope
              4%   Increased scope
              21% No response to scope
              5%   Faster completion
              51% Slower completion
              44% No response to completion rate
    14% Uncertain
    40% Probably not
    16% Definitely not
    4%   No response to question 12

Part III. Funding and other assistance

Commercialization of the results of an SBIR project normally requires additional developmental funding. Questions 13 and 14 address additional funding. Additional Developmental Funds include non-SBIR funds from federal or private sector sources, or from your own company, used for further development and/or commercialization of the technology developed during this Phase I project.

13. Have you received or invested any additional developmental funding in this Phase I? % based on 1,366.

APPENDIX C

25% Yes. Go to question 14.
72% No. Skip question 14 and submit the survey.
3% No response to question 13.

14. To date, what has been the approximate total additional developmental funding for the technology developed during this Phase I? (Enter numbers. If none, enter 0 [zero]; leave blank if unknown.)

Source	# Reporting that source	Developmental Funding (Average Funding)
a. Non-SBIR federal funds	79	**$72,697**
b. Private Investment		
(1) U.S. Venture Capital	13	**$4,114**
(2) Foreign investment	8	**$4,288**
(3) Other private equity	20	**$7,605**
(4) Other domestic private company	39	**$8,522**
c. Other sources		
(1) State or local governments	20	**$1,672**
(2) College or Universities	6	$293
d. Your own company (Including money you have borrowed)	149	**$21,548**
e. Personal funds of company owners	54	**$4,955**

Average Funding determined by dividing totals by 1,366 responders.

Appendix D

Case Studies

Airak, Inc.[1]

Nicholas S. Vonortas
Jeffrey Williams
The George Washington University

March 2005

THE COMPANY

Established in 1998, Airak is based in Ashburn, Virginia, and is currently focused on the development and commercialization of optical current and voltage transducers. To date, the firm has been granted a total of eight Phase I awards, five Phase II awards, and one Phase IIB supplemental award, from NASA, the National Science Foundation, and from the U.S. Department of Energy. The company's first major commercialized innovation, a fiber optic electrical current

---

[1]This case is based on primary material collected by Nicholas Vonortas and Jeffrey Williams during an interview with the owner and president of Airak, Inc., Mr. Paul Duncan. It is also based on preliminary research on the company carried out by the authors. We are indebted to Mr. Duncan for his willingness to participate and his generosity in offering both a wealth of information to cover the various aspects of the study and his broad experience with the SBIR program and with high technology development in the context of small business. All opinions in the document are solely those of Mr. Duncan. The authors are responsible for remaining mistakes and misconceptions.

transducer, has been picked up by the U.S. Navy and other industrial concerns, and completed products are to be shipped to these clients starting in early 2005.

Airak was originally conceived as a company that would focus on water quality monitoring innovations. In fact, their first SBIR award, a Phase I grant, was for a fiber optic remote monitoring system intended to measure dissolved oxygen levels in aquatic environments. Paul Duncan, the founder and owner, was working on his master's thesis at Virginia Polytechnic Institute at the time he developed the fiber optic dissolved oxygen sensor. He was familiar with the SBIR program through previous employment, and sought a Phase I award in order to get his company up and running. A Phase I award was granted by the DoE, and Airak was quickly formulated to take advantage of the opportunity. Without SBIR, the interviewee doubted that Airak would have been established.

While Airak's original concentration was intended to be water quality monitoring, the commercialization possibilities turned out to be insufficient to maintain the firm. Potential buyers were found, but never for more than a few of the dissolved oxygen sensors at one time. As the sensors are relatively inexpensive, a profit margin that would sustain the company could not be maintained. However, the inherent nature of fiber optics, most importantly the insulating effects of the glass inner structure, are well suited to electrical applications. The company began to develop several patents in the area of fiber optic electrical sensors and monitors. In the process, the firm has grown from only having one employee (the interviewee) to five at this point in time.

The company now focuses primarily on the electrical end of fiber optic sensing and monitoring solutions as there are a wider range of commercial possibilities in this area. While water quality monitoring innovations are not currently at the fore, the company goal is to one day be able to innovate and commercially develop their original product types.

FUNDING AND COMMERCIALIZATION

SBIR grants have been vital to Airak. Not only would the company most likely not have started had the grants not been awarded, but Airak has continued to use Phase I and Phase II awards to keep the company moving forward. Currently, SBIR awards make up 95 percent of Airak's revenue, totaling $3.7 million. The management, however, is well aware of the problems inherent in long-term reliance on SBIR funding: The successful transfer of products to the marketplace is perceived as the only way for an innovation company to survive and grow in the long term. To that end, Airak is expected to probably seek, at most, two more SBIR grants in the next year. Besides these, the company will focus all of its energies on the commercialization of the core products.

Prior to founding Airak, Paul Duncan was employed at another innovation firm. The business model for his previous employer was to rely almost exclusively on SBIR grants for long-term growth. At first, the model worked well.

SBIR grants are intended to help fledgling innovation companies get off the ground. During his tenure at his prior employer, the company grew from four to thirty-five employees.. Engineers were unable to follow-up on Phase II successes because funds were diverted to hiring new researchers who were to bring in new Phase I and Phase II awards, especially as ideas moved outside of the realm of the company's current knowledge base. The lack of focus, leverage between contracts, and large competition for internal resources became a significant issue for Duncan, and was a significant contributor to his departure.

With this earlier experience serving as an object lesson, Airak's management is reportedly fully devoted to the idea of commercializing products, and relying on future product revenues to grow the company. There is a gap here, though, that makes the transition to commercialization difficult for some firms. Government grants will typically help a company up to the production phase, but not into the marketing phase. Venture capitalists would rather invest in an innovation that has at least some proven track record. The hazards of this funding gap (the "valley of death"), in which an innovation may remain stuck if the company is unable, on its own, to market the product are well understood. To that end, the interviewee commented in favor of the ability to devote at least some fraction of the SBIR awards to marketing efforts, though he does understand the government's reluctance to become involved in the marketing matters of private companies.

External, nongovernment funding is considered a very important asset by Airak. One source of support is through the Virginia Tech Intellectual Property (VTIP) contract. While a graduate student at Virginia Tech, Airak's founder developed the dissolved oxygen sensor that was to earn the company its first SBIR award. VTIP, which owns rights to a related innovation, has licensed the product exclusively to Airak, in exchange for a 2-percent ownership of the company. As the company grows, VTIP will also receive a 2-percent share of the profits. Banks are also more likely to be willing to provide business loans to companies that have received a Phase II grant. The institutions reportedly recognize that a Phase II award is a good indicator of potential marketability of an innovation, and are thus more willing to provide funds that may be used for the commercialization of that product.

THE SBIR PROCESS

As stated earlier, Paul Duncan first became aware of the SBIR program through his previous employer. He was thus well aware of the application and granting procedures when he applied for the Phase I grant that was to serve as the launching platform for Airak.

Originally, Airak applied for a Phase I grant from pretty much all of the agencies, hoping that someone would find the dissolved oxygen sensor innovation worth further investigation. Now, however, the company focuses mainly on the Department of Energy and the Department of Defense, especially the Navy, for

SBIR grants. The fiber optic sensor technology in which the company specializes seems to best fit the needs of these two government agencies. And while the application process itself has not determined agency selection, the interviewee noted very pronounced differences among the various SBIR application processes. For example, the USDA, at least until recently, required several hard copies of the application and had no online functionality. As a contrast, the Department of Defense application process is completely online, and seems to be the most efficient in Airak's experience.[2]

As for the relationship between application cost and award funding, there is not a clear correlation. Even a Phase I award of $100,000 is insufficient to add another engineer or researcher to the payroll. First, few people will accept to work at one location for only six months, or until the Phase II application is ready to go. Second, even if there was a researcher willing to work for such a short stint, it is not likely that this individual could be adequately compensated from the Phase I award monies, as those funds must be divided among other employees, lab space, equipment, etc. Phase II, however, can potentially provide a robust grant, allowing for the long-term employment of one or more additional researchers and engineers. The application is longer for the Phase II awards, but, as noted by the interviewee, the time spent on the application is more than made up for by the potential for a large grant. The cost of applying for a Phase II award is the opportunity cost of missing out on an estimated one or two weeks of product work.

Airak has applied to one other federal program, NIST's Advanced Technology Program, for funding but has not been awarded a grant.

One major change to the SBIR program that Airak would support is additional commercialization assistance. Commercialization being the ultimately desired outcome for SBIR funding, the interviewee noted that for many DoD SBIR grants, there is a built-in buyer, thus decreasing the amount of energy that the awardee will need to put into commercialization. For other agencies, however, there is a much less robust built-in market.

Another possible change would be an effort to limit companies from using SBIR as their sole source of funding. Certainly, for those just getting off the ground, the SBIR program will be their major, if not only, source of revenue for some time. But established companies should have to clear some kind of benchmark—perhaps revenues from previous Phase II products, or at least an active commercialization effort—before they are allowed to apply for future awards. Ideally, this would force more innovations out of the workroom and into the marketplace. There should not be a limit to the overall number of SBIR awards to a company, assuming an effort is made to develop the products. Companies can innovate all the time, and all those innovations should still be encouraged.

---

[2]The Fastlane submission system of the U.S. National Science Foundation was also mentioned in very favorable terms.

Airak would certainly support an increase in submission frequency from once per year to at least twice per year, which is already a standard at the SBIR program of the U.S. Department of Defense. Technology moves quickly, and future market opportunities can come and go rapidly. If a company were to have a proposal ready one week after the deadline, then for most agencies, they would need to wait nearly a year for the application process to begin. Additionally, combining the SBIR/STTR applications at all agencies would be time and effort saving for the applicants.

Regarding the award time frame, it was indicated that the four years from Phase I application to Phase II completion is too long. If there were a way to speed up the process, or at least give the grantee the option to work more quickly, then that could aid in future commercialization efforts. For example, the National Science Foundation requires a business plan to be submitted with a Phase II application. Future market models, as predicted by the grantee, might be more accurate if the application occurred temporally closer to the commercialization phase.

Airak did face a problem with some of the proposal requirements for third-party participation. In their case, the third party wrote a letter of intent that worded the innovation support in too loose of a fashion, prompting the reviewing agency to turn down the application. It was felt that third parties should not have to submit this item, as it is unnecessary to the actual granting of the award.

Overall, Airak is very satisfied with the SBIR program. Without the grants, there would most likely not be an Airak today. Additionally, SBIR grants have allowed the company to maintain a steady growth, both in employees and funding, over the last five years. And just as is intended by the program, Airak will soon focus almost all of its energy on commercialization efforts, and limit its application for SBIR awards.

EXAMPLE INNOVATION FROM AIRAK

Airak's most successful innovation to date is the fiber optic electrical current transducer. Working on the Faraday effect, the sensor is designed to provide direct measurements of magnetic field intensity, current, and temperature in moderate to high-voltage environments. Potential markets include naval vessels, in which each ship is a self-contained power plant and electric grid, and civilian power companies, which need current and temperature monitors in environments such as switching stations, transformers, power lines, etc.

The electric current transducer earned Airak a Phase II grant from the DoE. Subsequently, the Navy awarded Airak a Phase IIB supplementary award. At present, Airak is set to deliver completed products to the Navy in 2005.

Atlantia Offshore Limited[3]

Grant C. Black
Indiana University South Bend

September 2004

THE COMPANY

Atlantia Offshore Limited began as a family-run small business located in Houston, Texas. Atlantia was founded in 1979 by husband and wife team, Joe and Pat Blanford. The company was created to provide full engineering services related to the design of shallow-water, low-tech platforms in the offshore oil industry. These fixed offshore minimal-production platforms were marketed to independent gas and oil companies operating in the Gulf of Mexico and North Sea. During the early years of the company, Joe Blanford was Atlantia's only fulltime employee. As projects required, additional consultants were hired to assist Mr. Blanford.

The location of the company in Houston, Texas, was believed to be vital to the success of the company. Houston is the center of the oil industry in the United States, with at least 80 percent of firms that design offshore platforms having operations there. Geographic proximity to these other firms provides close contact to most participants in the offshore industry. This proximity can reduce the costs of marketing to and working with other companies as well as providing access to a large, skilled labor pool familiar with the industry.

Atlantia's initial efforts were successful. The first shallow-water fixed platform designed by Atlantia was delivered in 1984, five years after the inception of the company. Since then Atlantia has designed more than 150 shallow-water fixed platforms. These platforms were constructed for approximately $2 million per platform. As Atlantia gained success in the shallow-water market, it began to recognize a need in the deep-water segment of the industry.

A recurring problem faced the industry: No viable cost-effective technologies existed to probe small fields for oil discoveries and pump them for production. Leases on these fields frequently expired with little or no activity on them. This motivated small oil enterprises to move into deeper water which was less competitive. Traditional deep-water platforms, however, are more expensive than shallow-water ones and were unviable for many small fields. To explore its potential for addressing this issue, Atlantia contacted Steve Kibbee. Kibbee, with a

---

[3]This case is based on an interview with Steve Kibbee, vice president of technology at Atlantia Offshore Limited. All opinions expressed in this report are solely those of Mr. Kibbee. Sincere thanks are expressed to Mr. Kibbee for his enthusiastic willingness to participate in this study.

APPENDIX D 171

history of deep-water research at large oil companies including British Petroleum, Texaco and Shell, was involved in early conceptual research on new discoveries. Kibbee joined Atlantia in 1990; the company employed approximately ten people at this time.

Beginning in the early 1990s, Atlantia began research that would shift the focus of the company away from shallow-water to deep-water engineering services. In 1992 the company received a patent for its SeaStar® tension leg platform (TLP).[4] By this time, Atlantia had expanded further, employing approximately 15 people. This new technology was designed to increase efficiency and reduce cost compared to existing technologies. The SeaStar® TLP can be adapted to wet- or dry-tree applications, is vertically moored to minimize the vertical heave and horizontal roll and pitch of the platform, has a monocolumn hull that can be built to any size, is modular by design so that production can take place at any fabrication facility or shipyard, and does not require the use of an expensive derrick barge for installation. These innovations in the SeaStar® TLP provided a new platform mechanism that substantially reduced the cost of developing small oil fields in deep water, allowing new development of fields that otherwise would not have been developed.

After the introduction of the SeaStar® TLP, Atlantia focused its marketing to relatively small European oil companies with undeveloped fields in the Gulf of Mexico. U.S. oil companies were initially hesitant to explore this new technology, so Atlantia quickly targeted its efforts to the more receptive European market. Sales soon followed. Atlantia completed installation of the first wet-tree SeaStar® TLP in 1998 for Agip's Moreth oil field. Trust in Atlantia's abilities based on previous shallow-water efforts was instrumental in securing this first sale of its innovative deep-water platform. Additional sales quickly followed. Agip purchased a second wet-tree SeaStar® TLP for its Allegheny field that was installed in 1999. Chevron Texaco purchased a wet-tree SeaStar® TLP for its Typhoon field that was installed in 2001.

In 2001 the French company, TotalFinalElf, selected the SeaStar® TLP for its Matterhorn field in the Gulf of Mexico. TotalFinalElf desired nine dry-tree TLPs to develop this field. The first dry-tree SeaStar® TLP, which was much larger than the previous platforms, was installed in 2003. This platform was also the first platform project in the Gulf of Mexico purchased under a fixed-fee basis. This lump-sum fee covered all costs including production, installation, and accompanying services.

To successfully manufacture and install these platforms, Atlantia needed a far broader range of skills than the current small number of employees could provide. Therefore, Atlantia rapidly expanded its staff to over 100—including additional managers, drafters and engineers—and heavily subcontracted the fab-

---

[4]For detailed information on the SeaStar® platform, visit Atlantia's Web site at <*http://www.atlantia.com*>.

rication process to other companies. Based on the experiences of these early installations, Atlantia has developed a flexible employment process. Atlantia maintains a core staff of approximately 15-20 employees necessary to optimally maintain its operations. Atlantia's core staff encompasses expertise in metocean, naval architecture, process facility, riser design and integration specialization, fabrication, and installation. Over 50 percent of Atlantia's employees hold a masters or doctoral degree. Employment is temporarily expanded as needed for large projects, with substantial use of subcontracting for areas of expertise and capabilities outside Atlantia's scope.

The SeaStar® technology has allowed Atlantia to provide a cost effective alternative for the development of relatively small oil fields in deep water. The SeaStar® TLP provides an efficient, low-cost mechanism to develop fields that would otherwise not be. The efficient design, production methods, and installation processes used for SeaStar® TLPs have also translated into the shortest time between project initiation and tapping the first oil from a new platform—only 21 months on average. After installation of its platforms, Atlantia continues its relationships with customers by providing custom services for the maintenance and efficient performance of its platforms. Atlantia believes these ongoing services strengthen important customer relationships and provide data to improve its research on existing and potential technologies.

Given the success of the SeaStar® technology, Atlantia found itself as a small company trying to sell a big product to big companies. While successful, Atlantia recognized that it could benefit from help in marketing on a grander scale in the international market. Atlantia's owners decided to sell the company to IHC Caland. IHC Caland is a Dutch holding company comprising companies broadly involved in marine technology. This group specializes in offshore oil services (including floating production systems), dredging, and shipping. Atlantia continues to operate relatively independently and has partnered well with sister company, Single Buoy Moorings (SBM), to market on an international basis by drawing on SBM's extensive marketing and sales operations. The strength of this merger is that Atlantia can benefit from the larger scale and scope of resources of the parent company, particularly access to finances, greater market presence, resources, and other technologies. Atlantia's expects an optimistic future for the company, with sales and overall growth of the company predicted to continue to rise. For instance, Atlantia currently has a contract to build and install the world's deepest platform (approximately 8,200 feet). It has recently completed its fifth platform and has proposals for three to four more new platforms in the near future.

DOE SBIR EXPERIENCE

Atlantia is an exception compared to most companies that have participated in the SBIR program. Atlantia pursued funding for only one research project—the

APPENDIX D

SeaStar® technology—which resulted in receiving one Phase I award and one Phase II award. This project was funded through the DoE SBIR program. Atlantia has not pursued SBIR funding from any other agencies.

Atlantia first became aware of the SBIR program in 1990. Steve Kibbee saw a brochure about SBIR at an oil industry trade show. After gaining interest in learning more about the program, Kibbee attended an SBIR informational conference. Given its involvement in the oil industry, Atlantia believed DoE would be the most appropriate possible fit for its research activities. Atlantia investigated the solicitations from DoE's SBIR program but believed none of the solicited topics easily fit its ideas developing around tension-leg platforms.[5] However, given the small size of the company and the steady business generated from its shallow-water products and services, Atlantia decided to apply to the SBIR program in an attempt to raise much needed financial support for its changing research direction.

With no background in the SBIR application process and no help from external sources, Kibbee virtually single-handedly prepared Atlantia's SBIR Phase I application, which was submitted in fall 1990. According to Kibbee, the company faced considerable difficulty trying to maintain its responsibilities and work on the SBIR application, let alone allocating the necessary time to work on the SBIR research once the project was funded. Kibbee estimates that he spent at least one calendar month of time preparing the Phase I proposal. Atlantia was awarded a Phase I grant for its SeaStar® technology and continued on this research through 1993 at the completion of its successive Phase II award that continued this research. It received the Phase II grant from DoE in 1991 as part of DoE's "early award" Phase II process. Kibbee found the Phase II proposal significantly more difficult and time consuming to prepare compared to the Phase I proposal. He estimates that approximately two calendar months of work time were devoted to this proposal. The transition between Phase I and Phase II went smoothly. Atlantia faced no significant disruption in its SeaStar® research between phases. A challenge, though, was the time needed to develop sufficient results from Phase I to use in justifying its Phase II proposal.

While time was needed to prepare proposals, especially given the company's inexperience with SBIR, Atlantia found participation in the SBIR program to be surprisingly straightforward and simpler than other types of research funding, such as acquiring funds in the private sector and other government programs. After receiving its SBIR awards, DoE sent checks for the amount of the awards and required little oversight. Atlantia benefited from this speedy, loose format; however, Kibbee recognizes that this lack of strict supervision could generate incentives for abuse of the SBIR program. Kibbee also strongly recommends that SBIR solicitations, at least from DoE, be more broadly defined to allow greater

---

[5] According to Kibbee, the solicitation topics focused on natural gas issues.

flexibility for firms to connect their research to DoE's interests.[6] As Kibbee pointedly remarked, Atlantia "almost did not do a proposal" because of feeling like its SeaStar® research did not seem to fit a published solicitation topic.

SBIR funding was vital to the success of the development of Atlantia's SeaStar® technology. Kibbee adamantly argues that the SeaStar® research would not have occurred without SBIR funding. At that time, Atlantia had approximately ten employees and the level of work from the shallow-water projects required considerable efforts from this small staff. Without financing through SBIR, Kibbee does not believe that Atlantia could have sufficiently diverted engineers involved in the developing SeaStar® research from other duties—and without their involvement this research would not have succeeded. Near 1990, SBIR funding contributed approximately 50 percent to Atlantia's research funding. The SBIR awards alone provided at least 1-2 months of revenues that were instrumental in keeping necessary staff employed.

Atlantia also benefited from commercialization-stage initiatives in DoE's SBIR program. Atlantia utilized Dawnbreaker, who is contracted to help SBIR firms at the commercialization stage. Dawnbreaker frequently helps firms develop business plans to commercialize their SBIR research, but Atlantia required other services as it tried to commercialize its SeaStar® technology. According to Kibbee, Dawnbreaker proved "tremendously helpful" in Atlantia's initial negotiations with British Borneo. This highlights the usefulness of providing flexible services to SBIR companies in trying to commercialize outcomes from their SBIR research.

SBIR OUTCOMES

Atlantia is an exemplary case of what the SBIR Program was envisioned to achieve. SBIR provided necessary short-term funding so that innovative research with significant commercial potential could be realized. The Department of Energy touts Atlantia as an SBIR success story, evidenced by its inclusion in the handful of companies listed on DoE's Web site as SBIR successes. Atlantia had minimal participation in the SBIR program—one Phase I and one Phase II award from DoE—but directly developed a commercial product from this SBIR sponsored research that quickly reaped substantial returns. According to Kibbee, before the SeaStar® technology created through its SBIR research, Atlantia earned approximately $10 million in revenues and employed approximately ten people. After the early commercialization of this technology, revenues jumped to $100 million and employment rose to over 100. This research rapidly transformed the direction of the company, shifting it away from its previous focus on shallow-water operations to deep-water TLPs. In addition, the SeaStar® technology developed

---

[6]Note that this perception of topic solicitation is based on Atlantia's limited experience with the SBIR Program in the early 1990s.

through SBIR research opened the door for innovations in Atlantia's shallow-water business. Revenues related to the first four SeaStar® TLPs alone exceeded $500 million for Atlantia. Moreover, the accompanying new services provided by Atlantia for these platform customers yield approximately $100,000 in increased revenues per platform. This can more than triple if unforeseen difficulties require more complex services. Along with ongoing services, Atlantia has developed new products related to its SeaStar® technology that also provide revenues. For instance, Atlantia has recently developed a $2 million riser that is nearly ready for its first order for the Matterhorn platform.

These financial returns not only have been captured by Atlantia but have also benefited the federal government. The U.S. government receives royalties from the oil production. The new platforms that tap into oil fields that would have otherwise not been developed yield real net gains in revenues for the U.S. government. It is estimated that the existing SeaStar® TLPs installed by Atlantia generate at least $100 million per year for the U.S government. These revenues from increased oil production will only climb as fields continue to be developed using Atlantia's SeaStar® technology.

The commercialized products and services resulting from Atlantia's SBIR research has yielded beneficial outcomes outside of revenues. Approximately four patents were granted directly from the SBIR research, including Atlantia's first successful patent. More than 100 foreign and U.S. patents have been granted to Atlantia based on its deep-water technologies that directly or indirectly stemmed from the initial SBIR research in the early 1990s. Atlantia has even patented designs to protect ideas for potential markets that have not developed yet.

Moreover, 20-30 professional papers have been written related to the SeaStar® technology and the existing TLPs that have been installed. Each new project generates a series of technical papers specifically related to it. These papers are frequently presented at oil industry conventions and other meetings, which disseminates new knowledge traced to the original SBIR research.

Atlantia received the Tibbetts Award in 1997 for its SBIR-related success. Kibbee believes that receiving this award generated considerable publicity for Atlantia, improving its marketing potential. Touting such SBIR successes, especially by the U.S. government, can be instrumental in providing a visible "stamp of approval" on companies such as Atlantia that can create a competitive advantage in their markets.

Creare, Inc.[7]

Philip E. Auerswald
George Mason University

August 2005

OVERVIEW

Creare, Inc. is a privately held engineering services company located in Hanover, NH. The company was founded in 1961 by Robert Dean, formerly a research director at Ingersoll Rand. It currently has a staff of 105 of whom 40 are engineers (27 PhDs) and 21 are technicians and machinists. A substantial percentage of the company's revenue is derived from the SBIR program. As of Fall 2004, Creare had received a total of 325 Phase I awards, 151 Phase II awards—more in the history of the program than all but two other firms.[8] While its focus is on engineering problem solving rather than the development of commercial products, since its founding it has been New Hampshire's version of Shockley Semiconductor, spawning a dozen spin-off firms employing over 1,500 people in the immediate region, with annual revenues reportedly in excess of $250 million.[9]

Creare's initial emphasis was on fluid mechanics, thermodynamics, and heat transfer research. For its first two decades its client base concentrated in the turbo-machinery and nuclear industries. In the 1980s the company expanded to energy, aerospace, cryogenics, and materials processing. Creare expertise spans many areas of engineering. Research at Creare now bridges diverse fields such as biomedical engineering and computational fluid and thermodynamics.

At any given point in time Creare's staff is involved in approximately 50 projects. Of the 40 engineers, 10-15 are active in publishing, external relations with clients, and participation in academic conferences. The company currently employs one MBA to manage administrative matters (though the company has

---

[7]This case is based primarily on primary material collected by Philip Auerswald during an interview at Creare, Inc. in Hanover, New Hampshire, on September 16, 2004 with Robert J. Kline-Schoder (Vice President, Principal Engineer), James J. Barry (Principal Engineer), Nabil A. Elkouh (Engineer). It is also based on preliminary research. A source on the early history of Creare was Philip Glouchevitch, "The Doctor of Spin-Off," *Valley News*, December 8, 1996, pp. E1 and E5. We are indebted to Creare, Inc., for their willingness to participate in the study and in offering both a wealth of information to cover the various aspects of the study and his broad experience with the SBIR program and with high technology in the context of small business. Views expressed are those of the authors, not of the National Academy of Sciences.

[8]The other two firms are Foster-Miller (recently sold, and no longer eligible for the SBIR program) and Physical Science, Inc.

[9]A list is given in the annex of this case study.

operated for long periods of time with no MBAs on staff). As Vice President and Principal Engineer Robert Kline Schoder states, "Those of us who are leading business development also lead the projects, and also publish. We wear a lot of hats."

The company's facilities comprise a small research campus, encompassing over 43,000 square feet of office, laboratory, shop, and library space. In addition to multipurpose labs Creare's facilities include a chemistry lab, a materials lab with a scanning electron microscope, a clean-room, an electronics lab, cryogenic test facilities, and outdoor test pads. On-site machine shops and computer facilities offer support services.

FIRM DEVELOPMENT

Founding and Growth

Creare's founder, Robert (Bob) Dean, earned his PhD in engineering (fluid/thermal dynamics) from MIT. He joined Ingersoll Rand as a director of research. Not finding the research work in a large corporation to his liking, he took an academic position at Dartmouth's Thayer School. Soon thereafter, he and two partners founded Creare. One of the two left soon after the company's founding; the other continued with the company. But for its first decade, Robert Dean was the motive force at Creare.

Engineer Nabil Elkouh relates that the company was originally established to "invent things, license the inventions, and make a lot of money that way." Technologies that would yield lucrative licensing deals proved to be difficult to find. The need to cover payroll led to a search for contract R&D work to cover expenses until the proverbial "golden eggs" started to hatch.

The culture of the company was strongly influenced by the personality of the founder, who was highly engaged in solving research and engineering problems, but not interesting in building a commercial company—indeed, it was precisely to avoid a "bottom line" preoccupation that he had left Ingersoll Rand. Thus, even the "golden eggs" that Bob Dean was focused on discovering were innovations to be licensed to other firms, not innovations for development at Creare.

As Elkouh observes "the philosophy was—even back then—that what a product business needs isn't what an R&D business needs. You're not going to be as creative as you can be if you're doing this to support the mother ship. . . . Products go through ebbs and flows and sometimes they need a lot of resources." Furthermore, Dean was a "small organization person," much more comfortable only in companies with a few dozen people than in a large corporation. A case in point: In 1968, Hypertherm was established as a subsidiary within Creare to develop and manufacture plasma-arc metal-cutting equipment. A year later Creare spun off Hypertherm. Today, with 500 employees, it is the world leader in this field.

By 1975, an internal division had developed within Creare. Where Dean, the founder, continued to be focused on the search for ideas with significant commercial potential, others at Creare preferred to maintain the scale and focus consistent with a contract research firm. The firm split, with Dean and some engineers leaving to start Creare Innovations. Creare Innovations endured for a decade, during which time it served as an incubator to three successful companies: Spectra, Verax, Creonics.

The partners who remained at Creare, Inc. instituted "policies of stability" that would deemphasize the search for "golden eggs"—ultimately including policies, described below, to make it easy for staff members to leave and start companies based upon Creare technologies.

The nuclear power industry became the major source of support for Creare. That changed quickly following the accident at Three Mile Island. At about the same time, the procurement situation with the federal government changed. Procurement reform made contracting with the federal government a far more elaborate and onerous process than it had been previously. As research funds from the nuclear industry disappeared and federal procurement contracts became less accessible to a firm of Creare's size, the company was suddenly pressured to seek new customers for its services.

In the wake of these changes came the SBIR program. The company's president at the time, Jim Block, had worked with the New Hampshire Senator Warren Rudman, a key congressional supporter of the original SBIR legislation. As a consequence, the company knew that SBIR was on its way. Creare was among the first firms to apply for, and to receive, an SBIR award.

Elkouh notes that "early in the program, small companies hadn't figured out how to use it. Departments hadn't figured out how to run the program." The management of the project was ad hoc. The award process was far less competitive than it is today." Emphasis on commercialization was minimal. Program managers defined topics according to whether or not they would represent an interesting technical challenge. There was little intention on the part of the agency to use the information "other than just as a report on the shelf."

IMPACTS

From the earliest stages of its involvement in the SBIR program, Creare has specialized in solving agency initiated problems. Many of these problems required multiple SBIR projects, and many years, to reach resolution. In most instances, the output of the project was simply knowledge gained—both by Creare employees directly, and as conveyed to the funding agency in a report. Impacts of the work were direct and indirect. As Elkouh states: "You're a piece in the government's bigger program. The Technical Program Officer learns about what you're doing. Other people in the community learn about what you're doing—both successes and failures. That can influence development of new programs."

Notwithstanding the general emphasis within the company on engineering problem solving without an eye to the market, the company has over thirty years generated a range of innovative outputs. The firm has 21 patents resulting from SBIR funded work.[10] Staff members have published dozens of papers. The firm has licensed technologies including high-torque threaded fasteners, a breast cancer surgery aid, corrosion preventative coverings, an electronic regulator for firefighters, and mass vaccination devices (pending). Products and services developed at Creare include thermal-fluid modeling and testing, miniature vacuum pumps, fluid dynamics simulation software, network software for data exchange, and the NCS Cryocooler used on the Hubble Space Telescope to restore the operation of the telescope's near-infrared imaging device.

In some cases, the company has developed technical capabilities that have remained latent for years until a problem arose for which those capabilities were required. The cryogenic cooler for the Hubble telescope is an example. The technologies that were required to build that cryogenic refrigerator started being developed in the early 80s as one of Creare's first SBIR projects. Over 20 years, Creare received over a dozen SBIR projects to develop the technologies that ultimately were used in the cryogenic cooler. Additionally, Creare has been awarded "Phase III" development funds from programmatic areas that were ten times the magnitude of the cumulative total of SBIR funds received for fundamental cryogenic refrigerator technology development. However, until the infrared imaging device on the Hubble telescope failed due to the unexpectedly rapid depletion of the solid nitrogen used to cool it, there had been no near-term application of the technologies that Creare had developed. The company has built five cryogenic cooler prototypes, and has been contacted by DoD primes and other large corporations seeking to have Creare custom build cryogenic coolers for their needs.[11]

Cooling systems for computers provide another example. The company worked intensively for a number of years in two-phase flow for the nuclear industry. This work branched into studies of two-phase flow in space—that is, a liquid-gas flow transferring heat under microgravity conditions. In the course of this work, the company developed a design manual for cooling systems based on this technology. The manual sold fifteen copies. As Elkouh observes, "there aren't that many people interested in two-phase flow in space." A Creare-developed computer modeling program for two-phase flows under variable gravity had a similar limited market. Ten years later, Creare received a call from a large semiconductor manufacturing company seeking new approaches to cooling its equipment because fans and air simply were not working any more. This led to a sequence of large industrial projects doing feasibility studies and design work

---

[10]Numbers as of Fall 2004.
[11]See NASA, 2002 (July/August). "Small Business/SBIR: NICMOS Cryocooler—Reactivating a Hubble Instrument," *Aerospace Technology Innovation*, 10(4):19-21. <http://ipp.nasa.gov/innovation/innovation104/6-smallbiz1.html>. See also <*http://www.nasatech.com/spinoff/spinoff2002/goddard.html*>.

to assist the client in evaluating different possible cooling systems, including two-phase approaches. The work covered the spectrum from putting together complete design methods—based on work performed under SBIR awards—to building experimental hardware. Most recently, NASA has contacted Creare with a renewed interest in the technology. From the agency standpoint, there is a benefit to Creare's relative stability as a small firm: They don't have to go back to square one to develop the technologies, if a need disappears and then arises again years later.

As academic research in the 1990s demonstrated the power of small firms as machines of job creation, the perception of the program changed. In the process, the relationship of perennial SBIR recipient firms such as Creare changed as well. These new modes of relationship, and some recommendations for the future, are described below.

Spin-off Companies

The success of the numerous companies that have spun off from Creare naturally leads to the question: Is fostering spin-offs an explicit part of the company's business model?

The answer is no to the extent that the company does not normally seek an equity stake in companies that it spins off. The primary reason has to do with the culture of Creare. Elkouh states that, as a rule, Creare has sought to inhibit firms as little as possible. "If you encumber them very much, they're going to fail. They are going to have a hard enough row to hoe to get themselves going. So, generally, we've tried to institute fairly minimal encumbrances on them. We've even licensed technology to companies who've spun off on relatively generous terms for them."

Does the intermittent drain of talent and technology from Creare due to the creation of spin-off firms create a challenge to the firm's partners? According to Kline-Schoder, no: "It has not happened all that often and when it has, opportunities for people who stay just expand. It's not cheap [to build a company] starting from scratch. So there's a barrier to people leaving and doing that. The other thing—in some sense, is that Creare is a lifestyle firm. Engineers are given a lot of freedom—a lot of autonomy in terms of things to work on. We think that Creare is a rather attractive place to work. So there's that barrier too."

ROLE OF THE SBIR PROGRAM

The founding of Creare pre-dated the start of the SBIR program by 20 years. However, SBIR came into being at an extremely opportune moment for the firm. It is very difficult to say whether or not the firm would have continued to exist without the program, but it is plain that the streamlined government procurement process for small business contracting ushered in by the SBIR program facilitated its sustainability and growth. In the intervening years, the SBIR program and technologies developed under the program have become the primary sources of revenue for the firm.

What accounts for the company's consistent success in winning SBIR awards? Kline-Schoder relates that "I've come across companies that have spun out of a university or a larger organization. I routinely receive calls—five years or more after I met these start ups—calling us and asking 'We were wondering, how have you guys been so successful? Can you tell us how do you do it?'"

As reported by the firm's staff members, Creare's rate of success in competitions where it has no prior experience with the technology or no prior relationship with the sponsor—"cold" proposals—is about the same as the overall average for the program. However, in domains where it has done prior work, the company's success rate is higher than that of the program overall. In some of these cases the author of the technical topic familiar with Creare's work may contact the firm to make them aware of the topic (this phenomenon is not unique to Creare).

Where the company has success with "cold proposals," it is often because the company successfully bridges disciplinary boundaries. In these instances, as Elkouh states, "we may have done something in one field. Someone in a different field needs something that's related to our previous work and we carry that experience over."

IMPROVING THE ADMINISTRATION OF THE SBIR PROGRAM

According to Creare's current staff members, the single most significant determinant of the Phase III potential of a project is the engagement of the author of the technical topic. Kline-Schoder states: "If your goal is to, at the end, have something that transitions (either commercially or to the government) having well written topics with authors who are energetic enough and know how to make that process happen. Oftentimes we see that you develop something, it works—it's great—and then the person on the other side doesn't know what to do. Even if you sat it on a table, the government wouldn't know how to buy it. There's no mechanism for them to actually buy it."

It is something of an irony that today, forty years after its founding, Creare is increasingly fulfilling the original ambitions of its founder: earning an increasing share of its revenue from the licensing of its technologies. Here, also, the active engagement of the topic author is critical. In one instance Elkouh worked with a Navy technical topic manager who saw the potential in a covering that had been developed at Creare with SBIR funds. This individual introduced him to over 300 people, and helped set up 100 presentations. That process led to Creare making a connection with a champion within a program area in the Navy who had the funds and was willing to seek a mechanism to buy the technology from Creare for the Navy's use.

However, even in this instance, concluding the license was not a simple matter. The appropriation made it into the budget—but that funding was still two years away. Elkouh: "The government funded the development of the technology because there was a need. Corrosion is the most pervasive thing that the Navy actually fights—a ship is a piece of metal sitting in salt water. There were reports from the fleet of people saying 'We want to cover our whole ship in this.' So now you have the people who use it say they want it, but who buys it? There is this vacuum right there—*who buys it*?"

With regard to contracting challenges, the SBIR program has largely solved the problem of a small business receiving R&D funds. From the standpoint of the staff interviewed at Creare, the contracting process directly related to the award is straightforward. What the SBIR program has not solved is the challenge of taking a technology developed under the SBIR program and finding the place within the agency, or the government, that could potentially purchase the technology.

Large corporations are no more willing to fund technology development than are government agencies. Kline-Schoder reports being approached by a large multinational interested in a technology that had been developed at Creare. The company offered to assist Creare with marketing and distribution once the technology had been fully developed into a product. However, the company was unwilling to offer any of the development funds required to get from a prototype to production.

Further obstacles to the commercial development of SBIR funded technology are clauses within the enabling legislation pertaining to technology transfer. Kline-Schoder: "FAR clauses were in existence before the SBIR program. They were inherited by the SBIR program, but they don't fit. For instance, they state that the government is entitled to a royalty-free license to any technology developed under SBIR. But there has never been a clear definition of what that means." In one instance Creare developed a coating of interest to a private company for use in a specific product. The federal government was perceived ultimately to be the major potential market for the product in question. The issue arose: Could the company pay a royalty to Creare for its technology, given that it would be prohibited from passing on the cost to the federal buyer? Contracting challenges related to the FAR clauses created a significant obstacle to the commercialization of the technology, even when two private entities were in agreement on its potential value. "We could potentially be sitting here now looking at fairly substantial licensing revenues from that product as would [the corporate partner] and it's not happening because of that IP issue."

A second issue pertaining to the intellectual property pertains to timing. As the clause is written, a company that invents something under an SBIR is obliged to disclose the invention to the government. Two years from the day that the company discloses, it must state whether or not it will seek a patent for the invention. However, the gap between the start of Phase I and the end of Phase II is most often longer than two years. So the SBIR-funded company is placed in the awkward position of being compelled to state whether or not it intends to seek a patent on a technology essentially before it is clear if the technology works. Pressure to disclose inventions has increased over time, as the commercial focus of the program has intensified. The time pressure is even more severe when Creare seeks to find the specific corporate partner who wants to use the technology in a product. The requirement also, importantly, precludes the SBIR-funded company from employing trade secrets as an approach to protecting its intellectual property—in certain contexts, a significant constraint. Kline-Schoder: "Patenting is not the only way to protect intellectual property. The way things are structured now, you don't have that choice. No matter what invention you disclose, you have to decide within two years whether or not to patent. If you don't patent, then the rights revert to the government." In this context, Creare has a much longer time horizon that most small companies.

The view expressed by the Creare staff members interviewed was that the size of awards is adequate for the scope of tasks expected. The variation in program administration among agencies is a strength of the program—although creating uniform reporting requirements for SBIR Phase III and commercialization data would significantly reduce the burdens on the company.

Finally, from an institutional standpoint, no substitutes exist for the SBIR program. Private firms often will not pay for the kind of development work funded by SBIR. Once the scale of a proposed project grows over $100,000, a private company will question the value of outsourcing the project. Lack of control is also a concern.

CONCLUSION

Creare appears to occupy a singular niche among SBIR funded companies. The company's forty year history as a small research firm is one characteristic that sets it apart from other SBIR funded firms. The many spin-offs it has produced is a second. However, from the standpoint of its ongoing success in the SBIR program and in providing corporate consulting services, Creare's most significant differentiating characteristic may be its range of expertise. The scope of the SBIR funded work at Creare is very broad. The reports of staff members suggest that the firm's competitive advantage relative to other small research firms is based to a significant extent on that breadth. "A lot of companies compartmentalize people," as Elkouh observes. "Everybody here is free to work on a variety of projects. At the end of the day, the companies I work with think that is where we bring the value." The same factor may account for the longevity of the firm. "We diversified internally by hiring people in different areas. That is when the cross-pollination happened." Areas come and go. Small product companies or small start-up companies focused in one area will struggle when the money disappears for whatever reason. Having evolved into a diversified research firm, Creare has endured.

CREARE—ANNEX

Sample of Independent Companies with Origins Linked to Creare

- Hypertherm, now the world's largest manufacturer of plasma cutting tools, was founded in 1968 to advance and market technology first developed at Creare. Hypertherm is consistently recognized as one of the most innovative and employee-friendly companies in New Hampshire.
- Creonics, founded in 1982, is now part of the Allen-Bradley division of Rockwell International. It develops and manufactures motion control systems for a wide variety of industrial processes.
- Spectra, a manufacturer of high-speed inkjet print heads and ink deposition systems (now a subsidiary of Markem Corporation) was formed in 1984 using sophisticated deposition technology originally developed at Creare.
- Creare's longstanding expertise in computational fluid dynamics (CFD) gave birth to a uniquely comprehensive suite of CFD software that is now marketed by Fluent (a subsidiary of Aavid Thermal Technologies, Inc.), a Creare spin-off company that was started in 1988.
- Mikros, founded in 1991, is a provider of precision micromachining services using advanced electric discharge machining technology initially developed at Creare.

APPENDIX D

Diversified Technologies, Inc.[12]

Philip E. Auerswald
George Mason University

September 2006

OVERVIEW

Diversified Technologies, Inc. (DTI) is a founder-owned engineering product and services company located outside of Boston in Bedford, Massachusetts. DTI was created in 1987 by Dr Marcel P. J. Gaudreau, previously the director of the Advanced Projects Group at MIT's Plasma Fusion Center, one of the founders of Applied Science and Technology, Inc. (ASTeX), in Woburn, Massachusetts. Created initially as a vehicle for Dr. Gaudreau's extramural research and consulting work, DTI had by the late 1990s developed into an industry leader in the application of solid-state devices to high-power, high-voltage opening and closing switches. Today DTI has annual revenues of approximately $11 million and is growing at approximately 20 percent per year. The company currently employs 63 people (57 full-time and six part-time) including 13 with doctoral degrees.

More than 90 percent of DTI's business is derived from its proprietary PowerMod™ solid-state switching technology, for which it has received two R&D 100 Awards—the first in 1997 and the second in 1999.[13] These solid-state modulators turn high power systems on and off at submicrosecond speeds in a repeatable and controllable way.[14] Potential applications for the technology exist in an array of markets from military radar to the commercial food industry. The company sells both to private and government customers through more than 50 different concurrent projects, ranging in size from multimillion dollar, multiple-year contracts to small, short-term contracts, on the order of $10,000. It also sells its products overseas, with representation by local distributors in France, Japan, and Korea.

---

[12]This case is based primarily on primary material collected by Philip Auerswald during an interview at Diversified Technologies Inc, on September 7, 2004, with Michael Kempkes, Vice President for Marketing. We are indebted to Diversified Technologies, Inc. for their willingness to participate in the study. Research assistance by Kirsten Apple is gratefully acknowledged. Views expressed are those of the authors, not of the National Academy of Sciences.

[13]The R&D 100 Awards as given on an annual basis by *R&D Magazine*. For additional information, including a description of the method by which award winners are selected, see <http://www.rdmag.com/awards.aspx>. Accessed September 26, 2006.

[14]From talk given by Dr. Marcel Gaudreau to the Aerospace and Electronic Systems Society on March 6, 2003.

From its inception, DTI has employed awards from the Small Business Innovation Research (SBIR) Program in a strategic manner. To date, DTI has been the recipient of 19 Phase II awards, with 15 of these from the Department of Energy and four from the Department of Defense. All but three of these awards have been received in the past seven years. The company's founder initially used the process of applying for SBIR awards as a way to engage graduate students in working through potential applications of academic research. In 1991, after many unsuccessful attempts, the company received its first award. That award was the basis for its proprietary solid-state switching technology. The company used later SBIR awards to demonstrate that the technology could be employed reliably in industrial settings, and to develop new technologies and markets.

FIRM DEVELOPMENT

1987-1996:
From a Diverse Consulting Practice to a Company with a Product Focus

From its creation in 1987 until it received its first SBIR award in 1991, DTI was something of a "virtual company," run out of the home of Dr. Marcel Gaudreau, the company's founder, while he worked full time as a researcher at MIT. At the time he started DTI, Gaudreau was already an experienced academic entrepreneur, having been involved in the founding of another company—Applied Science and Technology Inc. (ASTeX), which merged with MKS Instruments in January 2001. With DTI, Gaudreau was not seeking to create a high-growth firm, but rather a vehicle for his varied extramural consulting and research activities. A core activity of the company was the writing of SBIR proposals with graduate students, most of which were unsuccessful. Current DTI Vice President Michael Kempkes observes that "If the SBIRs hadn't existed, our company for sure would be a lot smaller. It probably would have remained in Marcel's house—more of a consulting company."

When DTI won its first SBIR in 1991, a Phase I award from the Department of Energy to develop a solid-state switching device, the company began to shift from wide-ranging consulting to tightly focused product development. The transition was slow and fitful. When DTI applied for a Phase II award from DoE to continue the solid-state switching project, the proposal was initially rejected. A second submission was, however, successful. By 1993, the company had completed the solid-state switching project and applied for a patent on the technology it developed.

The initial motivation for the solid-state switching project was purely academic. At the time the project was conceived, the concept appeared more like "science fiction" than a workable idea. Gaudreau was committed to proving that his concept could work in practice. The Department of Energy was interested in the project for possible application use in its fusion research program. As Kempkes describes: "He (Gaudreau) had a very good idea of where he thought the technology needed to go in the future. It was the combination of belief that this was something that was needed and that he could do it, and convincing the DoE to agree. That took a couple of steps. The tenacity was not so much 'we needed the money' as it was that Gaudreau was convinced this was a good idea and it should be done." During this time frame, Gaudreau remained employed at MIT.

Having proved the concept with the completion of the SBIR Phase II award, DTI began the initial marketing of products based on this technology, with limited success. The majority of DTI's work during this period remained consulting efforts for unique applications, performed mainly for acquaintances from MIT. One of these projects, a laser cutting system for a local manufacturer of photovoltaic arrays, was sufficiently large to require leasing of industrial space, and transition of the company into a full-time concern for Gaudreau and approximately a dozen of his students.

Among the key hires made at this time was that of Michael Kempkes, who joined as Vice President for Marketing in 1996. Kempkes faced the challenge of finding the element or elements in DTI's activities and history that could form the basis for a viable company. Most of the projects in which DTI was engaged were still consulting efforts or "one-off" prototypes in many different fields. DTI's patent for a solid-state switching device was one clear focal point within this otherwise confused set of activities. It was on this technology that Gaudreau and Kempkes decided to place their emphasis.

1996-1998:
Proving Technology Viability and Generating Market Acceptance

While the potential market for DTI's solid-state switches was small, it was well defined. The performance and size advantages of the technology were evident. Working against DTI was that the product would be used to power expensive vacuum tube amplifiers, such as klystrons and gyrotrons. Potential customers resisted adopting DTI's switches for the simple reason that they had not been proven to work reliably with the tubes, and a failure in the switches could damage or destroy the tubes themselves. As Kempkes notes, "people are nervous about doing anything but what the tube vendor thinks is acceptable. . . . We had customers, flat out tell us that they could not use the technology without the tube vendor's blessing. It took an awful lot of our time and effort to get over that hurdle."

The strategy forward gradually became self-evident. Again, DTI sought support from the SBIR Program. This time the objective of the project was not to prove the concept, but rather to work with a major RF-tube vendor to demonstrate that the technology was scalable for industrial use and superior in practice to available alternatives. With Phase I SBIR support from the Department of Energy, DTI partnered with Communications and Power Industries (CPI) in Palo Alto, a major manufacturer of RF vacuum tubes. DTI build a large switch that was incorporated into the CPI testing facility. Customers who visited CPI could see the PowerMod™ switch in use. Clearly, if the tube vendor used this technology itself, it must be acceptable. In this way the company overcame a major impediment to market acceptance.

A second major opportunity emerged out of a Navy SBIR, where two separate Phase I contracts were awarded to DTI and another small business in 1998. This SBIR required a switch for an advanced radar system under development at NRL that was simply not possible using conventional technology. After demonstrating a new version of DTI's solid-state switch in Phase I, DTI was awarded the sole Phase II contract from the Navy in 1999. This legitimized DTI's technology in the radar market, and led to DTI's first radar upgrade contract from the Navy later that same year.

Another breakthrough occurred in 1998, when DTI made a presentation on its technology at a conference hosted by the Stanford Linear Accelerator Center (SLAC). The focus of the conference was identifying the technologies and developments required to build a major new particle accelerator, the NLC. While the majority of this conference focused on conventional switching technologies, DTI was able to present its solid state technology to this group. The event drew considerable attention to DTI's technology. Not everyone was a fan—one senior manager stated that the NLC accelerator 'would go solid state over my dead body,' and DTI failed to win any SBIRs in 1998 related to this project. Over the next year, however, DTI's evangelism of the promise of solid state switching, combined with the delivery of the first system to CPI (only a few miles from SLAC), turned the tide towards solid state switching for the NLC. DTI won five Phase I SBIRs from DoE related to solid state developments for NLC applications in 1999, with four of these selected as Phase II efforts in 2000. This resulted in a considerable boost to DTI's R&D efforts, and significant growth for the company overall.

A further boost to DTI came with the receipt of the R&D 100 Awards in 1997 and 1999. While less important than the demonstration of the product achieved through the partnership with CPI, the initial radar efforts, and the reach achieved through the SLAC conference presentations, the R&D 100 Awards were nonetheless of value in generating additional attention for the company and validating its industry leadership.

1999-2004: Revenue Growth and Search for New Markets

With market acceptance of its core technology, DTI turned its attention first to consolidation of its revenue base, and then the exploring additional market opportunities.

As the company's focus has become increasingly well defined and its technical leadership better established, it has had increasing success with its applications to the SBIR Program. Of the 19 Phase II SBIR awards DTI has received since its founding, all but three have come since 1999. The company's success rate in Phase II proposals is close to 80 percent, well above the DoE average of approximately 40 percent. The company's experience with the program has allowed it to construct proposals that fit consistently within the SBIR Program's feasible funding range. Kempkes articulates the general lessons this way: "If the proposal is too far out, and you have no experience, you're never going to get it. And if too close, if it is a done deal, then you're also not going to get it." Proposals that fit between these two extremes—extending past successes in novel ways—have the best chance of being awarded. "There's a knee of the curve where this extends what you've already done, and you have experience in this area, but you haven't done the full thing. You've maybe done 50-60 percent of the work. That's where you have the best chance of getting the award."

This strategy of continual technology development and deepening of expertise in a core area of specialization has permitted DTI to find an entirely new and promising market application for its technology in the food services industry. An Army dual-use project at Ohio State University employed proprietary DTI technology and resulted in a prototype unit for removing bacteria from food at a commercial scale. As of 2006, the company was building a number of systems for both the food and water/wastewater processing markets based upon this new application of its switching technology.

The commercial applications of this technology have led to considerable commercial success. DTI's SBIR efforts have remained fairly constant since 1999, with approximately four to five Phase I and three to four Phase II efforts in process at any given time. The percentage of revenues DTI receives from SBIR funding, however, has decreased from nearly 75 percent of total revenues to less than 20 percent today.

KEY ISSUES AND LESSON LEARNED

When Loans Are Not an Option:
The SBIR Program as an Alternative to Equity Financing

As a consequence of his prior experience as a co-founder of a technology company, Gaudreau approached the founding of DTI with a commitment to building his new company without equity investments. As Kempkes relates: "From the outset [Gaudreau was determined that] DTI was going to be his, 100 percent. There would never be any equity. There would never be any investors There would never be any partners. He wanted to be able do with the company whatever he wanted to do, and not have to answer to anybody. And the company is still that way. We have shooed venture capitalists out the door." The preferences of the founder, Dr. Marcel Gaudreau, with regard to potential funding strategies have been a significant determinant of development of DTI. Gaudreau's determination to grow his company without equity investment, and the infeasibility of obtaining bank lending to fund the sort of high-risk, long time horizon work in which the company was and is engaged, left DTI with two feasible paths to funding: revenue from the sale of products and services, and competitive awards.

DTI has completed well over 1,000 consulting and development projects over two decades, including prototype development for DNA analysis, remote powering of robot devices, EMI analysis and control, underwater power and communication design, solar-powered vehicles, and turnkey product development and electronics manufacturing. Consulting services have focused on cost reduction, accelerated entry into new markets, and removal of technical barriers. Yet as Kempkes notes, a flexible upper limit exists to such corporate consulting contracts: Once the contract gets to be $100,000 or more, a company is likely to take the view that it can hire a full-time staff member internally to do the work, and doesn't require the outside engineering expertise. Consulting does not typically allow the consultant to retain intellectual property rights, making it difficult to leverage into other efforts. In recent years, consulting has been reduced to a very small fraction of the company's revenues.

While the company now has robust product sales, it could not sell a product until it had one. To develop its product, and demonstrate its application, competitive awards from the SBIR Program were essential.

The Importance of a Global Patent Strategy

Kempkes notes one potentially damaging error made by the company in the development of its PowerMod™ technology: only seeking patent protection in the United States. The company let pass a decade ago the one-year window for filing a patent overseas. As Kempkes notes: "At the time, no one thought to patent in Europe or Asia. No one knew if we should patent at all." The lack of an overseas patent appears to leave the company open to a competitive challenge from a potentially lower-cost non-U.S. producer. While a small group of U.S. and foreign competitors have emerged in the last two years, they have served to validate rather than erode DTI's market leadership. The competitive threat is mitigated by three factors that serve to strengthen DTI's competitive position:

- The product is both technologically complex and under constant development. The company undertakes a major reversion every year. As a consequence, reverse engineering or other forms of copying are not easily accomplished.
- The existing market for DTI's core technology is small—approximately $100 million in the United States, and double that worldwide. Given the potential for technical failure of an attempt to imitate a DTI product and the associated costs, the incentive to imitate is not very large. New applications, such as food processing, however, are emerging which may considerably increase the future size of this market.
- Increasingly, it is knowledge of the customer applications, rather than just the underlying technology, which provides DTI's competitive advantage. As

the first-mover in this area, DTI has more expertise and experience in applying solid state switching technology to a range of applications than any of the newer companies attempting to enter this market.

The Unresolved Challenge of Supporting Small Company Development of Commercial Products

While DTI was able to create a prototype of its technology with its first SBIR Phase I and Phase II awards—nearly a decade, and a number of additional SBIR awards, were required to advance the technology from a prototype to a stable product earning regular revenue. Moving a new technology into production generally takes $3-4 million—considerably more than the cost of developing the prototype. Kempkes notes, "the gulf between having the technology and having it in production in the market is probably bigger than the one between 'here's an idea, and here's a prototype.'" As the company enters new product areas, it once again faces the obstacles to commercialization of technologies developed with the assistance of SBIR funds—"Phase III" support in the parlance of the SBIR Program. It has only been since 1999, when the company's revenues passed $3 million annually, that DTI could afford to develop new products and markets from its product revenues—while continuing technology developments within the SBIR program.

While the SBIR Program is valuable in getting to the prototype stage, it cannot help with the refinement of products and associated processes required for mass-market production. Two SBIR proposals written by the company for product development were both declined for funding. Kempkes characterizes the reasoning on the part of reviewers and/or program managers as follows: "The technology exists, and if someone wants it they will fund it." Furthermore, the SBIR process is too lengthy. From proposal formulation to completion of Phase II can take four years. In many markets, that is too much time.

However, in DTI's experience as related by Kempkes, large corporations are hesitant to fund development of new production or process equipment unless they acquire full rights to the intellectual property. This reluctance creates an impenetrable barrier if the new technology could have potential industry-wide application. Understandably, in these cases, large companies would prefer for the development to be funded by others. Furthermore, even when the product is available, they would prefer not to depend on a sole-source supplier for a critical production component. As Kempkes put it: "The big companies do not want to spend the money. They don't want to develop the product; they just want to buy it once it is cheap." While understandable from the perspective of the large corporation, such reasoning creates a formidable set of obstacles to market entry—not only to develop the new product, but to await the emergence of a competitor in the same market. The government is often the only large customer willing to fund development of these technologies.

SUMMARY

The SBIR program has been essential to DTI's success and growth beyond the "garage" stage without external capitalization. The ability of the company to use the SBIR program to develop its technology, and to demonstrate it to potential customers in highly visible applications, provided the platform for the company to grow. Multiple awards from the SBIR program, in turn, allowed DTI both to broaden the application of its technology, and to achieve the 'critical mass' required to pursue commercialization and grow well beyond the SBIR program. Using the SBIR program in this way, however, required considerable perseverance on the part of the company's founder, as the transition from the garage to viable company took nearly a decade.

Eltron Research, Inc.[15]

Nicholas S. Vonortas
Jeffrey Williams
The George Washington University

March 2005

THE COMPANY

Eltron Research, Inc. undertakes basic and applied research in the areas of energy, chemical processing, and environmental processes and products. The company was founded in 1982 in Naperville, Illinois, and is presently based in Boulder, Colorado. Currently, there are approximately sixty employees, half of whom have Ph.D. degrees. Since 1983, Eltron has been the recipient of over 100 Phase I and Phase II SBIR awards from NASA, DoD, DoE, NSF, and NIH. This interview centered primarily on the company's DoE-related work.

There are six areas of focus within Eltron; catalytic membrane reactors, catalysis, fuel cells, materials research, electrolytic processes, and chemical sensors. Catalytic membrane reactors are used to isolate and extract gases, such as oxygen, hydrogen, and carbon dioxide, by simultaneously transporting electrons and ions through nonporous membranes. The membranes separate gases at low production costs via exothermic reactions, requiring no electrical or thermal energy inputs. Conversion of natural gas to synthesis gas is one application of this technology, yielding benefits of up to a 25-percent decrease in production costs, and eventual production of high-grade, crude-oil-competing liquids for $20 per barrel or less. Other applications, specific to oxygen separation, include coal gasification, H_2S removal, aromatic upgrading, atmospheric O_2 sequestering, alkanes to olefins, liquid fuel reforming, and environmental controls. Hydrogen and carbon dioxide each have associated commercializable applications, as well.

Catalysis has been a focus of Eltron Research since the inception of the company. Possible applications include converting synthesis gas into liquid fuels, inexpensive post-production removal of nitrogen oxides, low-temperature removal of volatile organic compounds from gaseous waste streams, shifting

---

[15]This case is based on primary material collected by Nicholas Vonortas and Jeffrey Williams during an interview with the founder and president of Eltron, Inc., Dr. Anthony Sammells at Boulder, Colorado, on February 23, 2005. It is also based on preliminary research on the company carried out by the authors. We are indebted to Dr Sammells for his willingness to participate and his generosity in offering both a wealth of information to cover the various aspects of the study and his broad experience with the SBIR program and with high technology development in the context of small business. All opinions in the document are solely those of Dr. Sammells. The authors are responsible for remaining mistakes and misconceptions.

carbon dioxide and hydrogen molecules to carbon monoxide and water molecules, advanced ignition catalysis for monopropellant fuels, devulcanizing tire rubber, and reduction in volatile organic compound emissions. Fuel cell technology revolves around using Eltron proprietary oxygen transport solid electrolytes to convert natural gas into electricity. Materials research is extensive, and is used to support the catalytic membrane reactors, catalysis, and fuel cell technologies. Electrolytic processes are being applied to disinfection and sterilization, organic contaminant destruction, biofouling removal in filtration units, and removal of nonindigenous species from ballast water. Chemical sensors focus on chemical defense, environmental monitoring, industrial hygiene and occupational safety, and process control applications.

Many of the innovations at Eltron Research are geared towards applications in the energy processing industry. An example of a suite of such innovations, centered on membrane separation technology, will be discussed in a later section.

SBIR AND ELTRON RESEARCH

Eltron Research received its first SBIR from the Department of Energy in 1983, about one and a half years following its founding. The firm owner first heard of SBIR at a conference held in Illinois, and Eltron Research was one of two recipients in the state during SBIR's first year. While SBIR was not the reason for founding Eltron, SBIR grants have been instrumental in the development of the company, especially by allowing it to build a much broader and deeper technology base in a more thorough manner than would have been otherwise possible. SBIR awards have served as seed money to begin new projects, with multiple awards in complex areas allowing the building up of capabilities in more than one complementary technology. Those projects have in turn attracted significant research funding from the private sector.

Eltron's principal source of financing is investment by private energy concerns, such as oil companies, and to a lesser extent government contracts. SBIR grants, while instrumental in the sense mentioned above—seed money allowing the establishment of new complex projects and continuing until attracting private funding becomes a possibility—have never been the principal form of funding for the company. The company has intentionally shunned private venture capital in its effort to maintain full control, escape stringent ownership conditions imposed by venture capitalists, and continue on the long-term trajectory of a cutting-edge organization.

The primary commercialization strategy of the firm is to create partnerships with users of its technology whereby it provides licenses for a fee, but also frequently for a percentage of the purchasing firm's earnings that are related to the innovation. As such, the relationships tend to be more collaborative than a straight sale.

In summary, a long-term strategy appears in which support from multiple sponsors—including SBIR awards—allows the company to build up internal capabilities in new complex technology areas while keeping full control of the operation. As the base grows, the firm is able to apply its novel technologies and applications to different situations, providing prospective customers with a wider range of solutions and solution techniques, raising the chances for commercial application (take-up). In such an event, the company stays with the customer long enough for a successful transfer of earnings returns from partnerships established around technology licensing agreements.

SPECIFIC COMMERCIAL APPLICATION

Eltron Research has received numerous grants in a wide variety of fields. One of the more successful resulted in the now-maturing technology of oxygen separation membranes. This technology isolates oxygen and nitrogen through the use of catalysis membranes. Development of the innovation began with a Department of Energy SBIR award in 1995. Eventually, two Phase II awards were associated with this technology, which has now generated a sizable income stream for the company.

Oxygen separation from the atmosphere occurs when air passes through a specially designed mixed ionic and electronic conducting membrane. Membranes are three-layered constructs; on the air side is a reduction catalyst, followed by a mixed conducting membrane, followed by a partial oxidation catalyst on the output side. This technology is a replacement for cryogenic oxygen separation which is very energy intensive.

The membrane conducts electrons and ions at high temperature. Eltron uses internally designed, low-cost thin film membranes to increase the flux for oxygen separation. Oxygen is available for a wide variety of industrial uses.

Processes such as coal gasification rely on the use of pure oxygen. Coal gasification has the ability to use coal more efficiently than traditional firing processes; 42-50 percent utilization versus 34 percent utilization in alternative processes. In addition, by-product emissions are significantly lower with coal gasification. Currently, oxygen is the third largest volume chemical used in the United States in general, at approximately 800 billion cubic feet in 2002. Use of oxygen is expected to grow, adding $1 billion to the oxygen market between 2002 and 2007. Eltron has licensed this innovation to the energy industry for further development.

The interview made apparent that Eltron is keenly focused on technology commercialization. Not surprisingly, patents are a key aspect of the firm's commercialization strategy. For example, it owns twenty-six core patents for the oxygen separation membrane technology. The typical procedure is for Eltron to cover the costs for U.S. patents of its technologies, while its licensing partners, which frequently are large energy concerns with extensive international interests, file foreign patents. This is considered especially helpful as the patent filing and maintenance fees, as well as other financial requirements for being able to credibly defend intellectual property rights in foreign markets tend to be prohibitively high for small companies.[16]

The company collaborates with universities and research laboratories fairly infrequently. The reason seems to be differences in culture towards the marketplace.

VIEWS ON SBIR

Eltron's management is, on the whole, happy with the SBIR award process. The company has received multiple awards, both Phase I and Phase II, from nearly all SBIR-granting agencies. The programs at the Department of Defense and the Department of Energy are seen as the best ones to work with, while the National Science Foundation can be more difficult. For example, while the DoD proposal submission process was viewed as streamlined, Eltron has reportedly encountered difficulties with the NSF Fastlane process. In addition, limitations on the number of allowable proposals per company per year by the NSF are an issue.

While there does not seem to be a clear separation of focus in the type of technological knowledge supported by the SBIR awarding agencies, the NSF is perceived as pursuing proposals for more basic and long-term research, while the DoD may be at the other end, looking for much more specific, solution-oriented proposals. The interviewee stressed the importance of the fact that real technological progress often occurs when a bunch of complementary innovations are pursued at the same time.[17] To the extent that this happens in a company, featuring funding from multiple sources to pursue related aspects of research, it complicates attribution of exact return streams to specific innovations and specific research projects, especially when those involve more basic, long-term research. Such questions were reported to be tricky and difficult to answer, especially when one tries to take into account the social value of an innovation on top of its value in terms of revenue to the producing organization.[18]

The management of Eltron does not see a need for much change in the SBIR application process, in general. Cost and time spent on the applications are not major concerns at Eltron as the firm has acquired experience in the process and

---

[16]An interesting observation was made during the interview. In the early stages of the company, scientific publications were reportedly a focus. This is expected given the high concentration of doctorates in the company's roster of employees. More recently, however, the emphasis on such channels of free knowledge dissemination has been downgraded in favor of formal channels of knowledge ownership affording the company the ability to commercialize ideas.

[17]Note that this issue has been amply stressed in the extant business and economic literature on technological advance ("network spillovers").

[18]This refers to the difference between social rates of return and private rates of return in economic jargon.

APPENDIX D   *197*

always has a backlog of prospective successful ideas that have not yet found alternative sources of support. The system is viewed as being fair in terms of awards granted. While the interviewee is aware of the extent of subjectivity involved in selecting proposals for support, this is understood as a general weakness of the peer-review system as a whole that one has to live with, and not as a specific weakness of the SBIR award process.

Regarding the actual funding process, the grants usually arrive in a timely manner. DoE was considered to be the most prompt agency in that respect, with payments occurring fairly rapidly. Besides that agency, the average time lag between invoicing and receiving the funds was estimated to be between 60 and 90 days. It was observed that this delay should be looked at and shrunk as much as possible, since it adversely affects smaller companies with intense cash flow problems. A possible increase in funding was considered appropriate given that the funding ceilings have not changed for a long time. For example, $75,000 for a Phase I award does not go as far as it used to. It is reportedly becoming very difficult to fully fund Phase I activities solely from a Phase I award.

Cash flow is, of course, a major issue for small innovative firms. Many projects need large amounts of money up front, not only for salaries but also for complex instrumentation. It was at this point that the interviewee made a very important, in our view, observation. This has to do with the perceived unevenness of competition between academic off-shoots and independent small, private companies for the same award in areas requiring significant scientific instrumentation. At a university setting, it was argued, a grantee is able to hire post-docs for little money and often has access to university equipment at no extra charge. Private firms, on the other hand, need to purchase or rent all of their equipment and must hire researchers at a significantly higher salary range. The potential disparity in the ability to compete for a grant should be taken into consideration by policy decision makers.

One would think from the above that a dollar in grant funding going to a company with a strong academic affiliation would have a magnified impact due to leveraged resources. Not necessarily so, according to the interviewee, because of vastly different incentives. The incentive structure of an academic environment may be such that it does not foster the most efficient use of resources when the ultimate purpose of the funding program is commercialization. In the private sector, according to the interviewee, researchers work in earnest under the market discipline: They may lose their jobs in case of a commercial flop. In an academic environment, however, if the innovation is not commercialized, there are few or no repercussions. As a consequence, funding that goes to academic grantees was perceived to stand a lower chance of generating a commercializable good that contributes directly to the economy.

The arguments above raise important points about the SBIR program that may deserve to be reconsidered:

- Is there a level playing field between independent, small private companies and their counterparts with strong academic affiliations at the proposal stage?
- Is the rate of commercialization of technologies from SBIR awards between these two kinds of companies significantly different?
- If so, should the phenomenon of multiple awards by a single company be looked at differently on the basis of the type of company involved?

With respect to the latter question, the interviewee emphasized the forced-efficiency nature of working in a private firm that contributes to a relatively high degree of successful innovation research. In his words, the company owner is "fueled by insecurity and self-doubt." Since reliance on a single innovation to carry the firm is unrealistic, the firm must constantly build on old technologies and strive to create new, commercially viable ones.

It should be stressed at this point that the question of multiple awards also links directly to the earlier observation of the complexity of the technological area pursued by a company.[19] The more complex the technology is, the larger the number of complementary pieces that need to be advanced in order for the technology to be of practical use. This has implications for the number of SBIR awards a company may need before reaching the point of commercialization of what may widely be perceived as a single innovation.

Finally, it was observed that procuring risk financing for early stage research outside of an SBIR-style grant is difficult. Commercial banks ask for collateral to back up expectations of future knowledge and technology, while venture capitalists demand at least partial control and look for near-term results. SBIR awards fill the vacuum. It was considered desirable for Congress to raise the SBIR earmark above the current 2.5 percent of the R&D budgets of research-intensive federal agencies, thus allowing the SBIR program to increase the grant amounts. Characteristically, the interviewee observed that 'the SBIR represents the most efficient use of R&D money in this country. There is an incredible amount of activity under this program."

---

[19]Technological complexity in this sense does not imply high tech. Rather complexity relates to the different pieces necessary for a technology to function. Complex technologies relate to products or processes that cannot be understood in full detail by an individual expert. In contrast, simple technologies, which may be quite advanced, can be fully understood by an expert (e.g., chemical, pharmaceutical compounds). See Don Kash and Robert Rycroft, "Technology policy in the 21st century: How will we adapt to complexity?" *Science and Public Policy*, 25(2):70-86, 1998.

IPIX, Inc.[20]

Nicholas S. Vonortas
Jeffrey Williams
The George Washington University

July 2005

THE COMPANY

IPIX markets proprietary technology for the creation and manipulation of still and video imaging for government and private industry clients. The nineteen year-old company is publicly traded on NASDAQ, and currently employs approximately sixty-five people at its headquarters in Oak Ridge, Tennessee, and offices in Reston, Virginia. During the past couple of decades, IPIX has received several Phase I and Phase II awards but no Phase IIB grants. IPIX now relies on commercially sold, mature iterations of its immersive imaging technology for nearly 100 percent of company income.

In 1986, a group of researchers at Oak Ridge Laboratories got together to establish the company that would later be known as IPIX. The firm came into being with the assistance of SBIR funding from the Department of Energy, granted for the development of a specialized camera to withstand the highly demanding environment that would theoretically exist inside of a fusion reactor.

In the early days SBIR grants kept the company alive and allowed it to maintain operations. According to Dr. Egnal, however, SBIR funds earned have not been intrinsic to the growth of the company since then. Instead, the firm has relied on external investors to supply the vast majority of growth capital. Private investment has been the driving engine for expansion through most of the company's history. For example, at one point in the late 1990s, the company

---

[20]This case is based on primary material collected by Nicholas Vonortas and Jeffrey Williams during an interview with the Chief Technology Officer of IPIX, Inc., Mr. Geoffrey Egnal. It is also based on preliminary research on the company carried out by the authors. As a newcomer to the company, Mr. Egnal expressed his reservations about being able to accurately reproduce historical details regarding the role of the SBIR program in the relatively long history of the company. The interviewers accepted his concerns and continued the interview given: (a) that he has a good grasp of the more recent developments and could access company SBIR records easily; and (b) the fact that the company has gone through important gyrations in the past 5-10 years as a result of market conditions with significant employee turnover which makes it very difficult to locate other sources of organizational memory. We are indebted to Mr. Egnal for his willingness to participate and generosity in offering both a wealth of information to cover the various aspects of the study and his broad experience with the SBIR program and with high technology development in the context of small business. All opinions in the document are solely those of Mr. Egnal. The authors are responsible for remaining mistakes and misconceptions.

employed nearly 1,200 people and was worth about $2.7 billion in the market. In reaching this stage, the firm mirrored the growth of the Internet, changing its name from TeleRobotics to OmniView, to Interactive Pictures Corporation, to Internet Pictures Corporation, and finally to IPIX. As the dot.com industry suffered, so did IPIX. Today, the firm, now with only 65 employees, is once again eligible for SBIR grants.

Having only arrived at IPIX approximately four months prior to the interview, Mr. Egnal was not certain he could draw the relationship between earning SBIR awards and the firm's ability to secure other financing. However, based on his experience with the SBIR program at previous companies, Mr. Egnal could confidently claim that the awards typically lend credibility to a firm and also enable the firm to be eligible for, and earn, other federal grants and contracts. In terms of research and development (R&D) funding, Mr. Egnal believes that the combination of SBIR awards and extensive commercialization has, historically, allowed IPIX to obtain private funding, and that some combination of these factors will continue to work to the firm's advantage in the future.

EXAMPLE OF AN SBIR RELATED INNOVATION

A combination of SBIR grants, other federal contracts, and private funding has enabled IPIX to expand its technological base. The firm's current product lines, made up of the software and hardware used to create and process still and video immersive images, stems directly from an innovation related to early DoE and NASA SBIR grants obtained in the late 1980s and early 1990s. That innovation was a software technique called fish-eye image de-warping. IPIX products, both still and video, combine in real time several images taken by fish-eye lenses to provide a 360-degree view from a single viewpoint. The SBIR grants assisted in the development of a technology that allowed the translation of the warped fish-eye image into a conventional, flat-perspective image. The flat-perspective images then appear in a 360-degree panoramic format, which is considerably easier to view as it lacks the distortion caused by the fish-eye lens.

All of IPIX's current products utilize the de-warping technology in order to render easily discernable images. Currently, all of these products are available in the commercial market and are produced in-house. IPIX manages all of the marketing and sales of its own product lines. The original de-warping technology, which was patented by the firm, has also formed the basis of more than twenty additional patents issued to the company. While the company has published several scientific papers, publication is not actively pursued as a knowledge dissemination strategy.

Technology utilizing, and derived from, the SBIR-funded de-warping innovation is being applied in a number of different venues. Both the still and video imaging technologies provide users with a 360-degree view from a single point of reference. Applications include security cameras, which involves a single camera capable of viewing an entire room, and real-time video monitoring. De-warping technology appears on real estate Web sites, allowing customers to view room interiors and exteriors as if they were actually standing in that position. Benefits to users of the technology include lowered costs due to decreased maintenance as there are no moving parts to the camera systems, and increased efficiency due to decreased blind spots and increased compatibilities associated with using digital images.

Dr. Egnal felt unable to answer questions regarding future market projection for IPIX products because of the publicly traded nature of the company.

IMPRESSIONS REGARDING THE SBIR PROCESS

Being new to the firm, Mr. Egnal was not familiar with all of the details regarding IPIX's experience with the SBIR process. However, by drawing on SBIR exposure from previous job assignments and learned knowledge of IPIX's history, he was able to provide a good amount of background information. Geography and familiarity with the needs of the agency were considered to be important factors in initiating the original award: The company founders were all employed at the Oak Ridge National Laboratories, one of the large research laboratories of the Department of Energy. Mr. Egnal also believed that, at least during the first few years of the firm, personal and professional networking among IPIX and former colleagues at Oak Ridge must have influenced extensively the firm's focus on DoE as the target application agency for SBIR awards.

In Mr. Egnal's experience, there are marked differences among the SBIR granting agencies. DARPA, for example, allows the program manager to review the SBIR applications, while other agency applications will go directly to a reviewing committee. DARPA program managers are more empowered than those in other agencies, and can be quite influential in selecting projects for funding. Agencies also differ in what exactly they want from SBIR applicants, influenced in this respect by their mandates. According to Mr. Egnal, the DoD is ultimately looking for finished products that it can procure, DARPA shifts between less focused research and specific product-oriented R&D projects, and NSF is more likely to support innovative ideas grounded in basic research. Other agencies fall in between. As such, IPIX alters the choice of projects selected for the SBIR application process depending on the target agency. Innovations closer to implementation may be more appropriate for DoD or DARPA, while other agencies might be more appropriate for innovations still in their early stages. Echoing other interviewees of ours, Mr. Egnal stressed the importance of business/professional networking for success: If a firm has connections within an SBIR-granting agency, meaning that agency managers are aware of the firm's capabilities and expertise, then there is occasionally a chance to influence the scope of a specific call (for proposals) when it is judged that agency needs will be better met. In such cases, there is an opportunity for that firm to apply for that round of funding.

Overall, Mr. Egnal felt that the SBIR application process is satisfactory and requires no major changes. Feedback was judged to be quite good and informative, telling the applying firm exactly what is missing or lacking in the application or innovation. Costs associated with application are more than made up for by the benefits, though he did caution that firms with no connections will find the process less easy to navigate and more cumbersome than in firms with an established network. Requirements of third-party investing for applying for Phase IIB funding are also not a problem for IPIX.[21] However, Mr. Egnal did say that for smaller firms, of between one and ten employees, attracting outside investors may stretch a firm's resource limits.

Regarding topic specification, the interviewee clearly recognized that each agency has its own mission, and thus, it is difficult to generalize as to whether topics should be tighter or more open. Nevertheless, the preference was for lesser definition of the topic specifications in order to increase transparency, ease topic selection for applying firms, and allowing easier entry by smaller participants. Regarding transparency, this preference goes along with the suspicion that, occasionally, very tight definition of specifications could imply a lock on the grant by a particular applicant whose orientation and capabilities are better reflected in the specifications.[22] Obviously, powerful program managers that exercise strong control over their programs could more easily impose tight specifications.

IPIX has applied for, and received awards from, federal R&D programs outside of the SBIR process. The firm has experienced differences among programs, which it assumes are natural given the different program goals. For example, ATP awards tend to have a better ratio of funding per proposal than does SBIR, but the SBIR process is more geared towards encouraging commercialization. And broad area announcement programs may have larger award amounts, but they are also geared towards satisfying an individual agency's needs rather than establishing commercial potential for an innovation.

Concerning funding, Mr. Egnal sees little problem with the funding structure for the SBIR program. In his experience, funding delays, if existent, have been inconsequential. He also felt that the level of funding and number of awards are at a good ratio, and does not see a need for change towards a system that grants fewer awards for larger amounts of money. The current funding and award ratio, he said, allows for a wide variety of topics to be presented by the agencies while also encouraging firms to keep applying for SBIR grants. A large increase in funding might lead to complacency on the part of the applying firms, as they would be able to stretch the award over a longer period of time. Nonetheless, Mr. Egnal thought that a small increase in Phase II awards, perhaps up to $1 mil-

---

[21]Recall that the firm has no Phase IIB awards, though.

[22]The interviewing team has met this opinion in several of its interviews with other companies. It must be emphasized that this is not an issue just with SBIR awards, but with all calls for proposals anywhere in the world. For example, it has been an important concern in the R&D programs of the European Union.

lion, would help defray higher costs of operating in today's commercial climate while not allowing the firm to become dependent on the award monies.

Overall, Mr. Egnal felt that the SBIR program is strong. There are very few methods by which small companies can raise funding without giving up equity or control. SBIR funds make it possible for those companies to take risks on an innovation without worrying too much about losing ownership of the firm itself. The program's emphasis on commercialization is also seen as a strong positive quality, as it really supports the long-term stability and growth of the grantees. Mr. Egnal expressed his belief that, due in large part to the push for commercialization, the process enjoys a very high metric of success. In turn, this metric of success creates social benefits: Innovations that would otherwise remain unrealized can be developed into commercializable products. Along those same lines, the technology produced as a result of an award is a benefit in and of itself to the innovating company, as the firm is able to develop that technology and then apply it to new, company-growing technologies and innovations.

Finally, Mr. Egnal reiterated some features of the SBIR process possibly amenable to changes. As mentioned earlier, he felt that it is of the utmost importance to maintain transparency in the generation of RFP's, and in the award process itself, as this would help ensure a level playing field, especially for the smallest or newest firms that do not have established networking capabilities. The influence of networking and the benefits associated with being familiar with the government contracting system and having personal contacts within granting agencies have been mentioned by several firms over the course of the DoE SBIR investigations, and Mr. Egnal echoed these thoughts by mentioning possible disadvantages for unconnected firms. A second area that requires attention in the SBIR process, in his opinion, would be the time gap between the Phase I and Phase II rounds, as it is difficult for some firms to maintain employees or facilities during the intervening time. He closed by stressing that these issues are overshadowed by the benefits, both to society and the individual firm, associated with the SBIR program.

NanoSonic, Inc.[23]

Nicholas S. Vonortas
Jeffrey Williams
The George Washington University

July 2005

THE COMPANY

NanoSonic, Inc, is a privately owned research and development concern based in Blacksburg, Virginia, with offices in Seattle, Washington, and Dayton, Ohio. The research focus of the firm is in the area of advanced materials and nanotechnology applications. A principal goal of the firm is to scale-up nanoscale properties, that is, transitioning nanoscale materials into macroscale materials that retain the characteristics of the original product. Such a process allows the macroscale material to retain the unique physical properties of its nanoscale base, properties that cannot easily be generated in macroscale materials.

The firm was founded in 1998 with three part-time employees, and has since grown to fifty-six employees and has received nearly fifty Phase I and over twenty Phase II SBIR awards from NASA, the Department of Energy (DoE), the Department of Defense (DoD), and the National Science Foundation (NSF), and has one Phase IIB with the NSF. Metal Rubber™, a highly plastic and highly electrically conductive nanocomposite film, is one of NanoSonic's premier products and is based on one of the firm's original SBIR-funded innovations, the modified self-assembly process. The firm has generated several patents as a result of SBIR/STTR funded research.

In 1998, Dr. Claus, a professor at Virginia Tech, was working on a materials contract with an entity from outside of the university. In order to pursue work on this project, Dr. Claus formed NanoSonic, an independent company outside of the purview of Virginia Tech. Even though the SBIR program was not the reason for establishing the company, it was soon to play a very important part in the company's development. Just after the company's creation, NanoSonic became aware of an Air Force solicitation, and the company responded, eventually earn-

---

[23]This case is based on primary material collected by Nicholas Vonortas and Jeffrey Williams during an interview with the president of NanoSonic, Inc., Dr. Richard Claus, and the Vice President of Business Development, Dr. Jennifer Lalli. It is also based on preliminary research on the company carried out by the authors. We are indebted to Dr. Claus and Dr. Lalli for their willingness to participate and generosity in offering both a wealth of information to cover the various aspects of the study and their broad experience with the SBIR program and with high technology development in the context of small business. All opinions in the document are solely those of Dr. Claus and Dr. Lalli. The authors are responsible for remaining mistakes and misconceptions.

ing a Phase I award from the DoD in 1998. Since that time, SBIR awards, which account for approximately 80 percent of the firm's current revenue, have been important for the continued growth of NanoSonic.

Typically, the company reinvests a portion of internal funds into product development. This also includes patent acquisition from the U.S. Patent and Trademark Office. In the past couple of years, as the company transitions from an R&D concern to a fully developed company, NanoSonic has also accelerated its direct marketing efforts with reportedly successful results.

FUNDING AND COMMERCIALIZATION

An important facet of SBIR applications and awards is reportedly a firm's familiarity with the government contracting process as a whole. Accordingly, it was indicated that an important factor to NanoSonic's success in obtaining awards has been Dr. Claus' prior exposure to government contracting, as that allowed the firm to navigate the process more confidently. His government security clearance and broad familiarity with how agencies, such as DoD, promote and solicit SBIR projects has seemingly been helpful, as well. Prior working experience with different federal agencies has allowed NanoSonic to anticipate agency needs and to know where to look for commercialization possibilities.

NanoSonic has experienced differences among SBIR-funding agencies in terms of efficiencies in working with scientists and engineers. Expectations can also differ about how materials will eventually get into the marketplace. Agencies with their own labs and intramural R&D tend to be more technologically focused. Further specifics were not provided by the interviewees—apart from the fact that some agencies, reflecting their mandates, are more science-oriented and some more application-oriented. But DARPA was singled out as the as the best agency with which to work for scientists and engineers.

Success in the SBIR program has also allowed NanoSonic to access additional sources of funding. Most importantly, it was felt that awards allow smaller companies to get inside the doors of the larger government contractors. An SBIR award is helpful because it raises flags for the prime contractor and instills confidence that the smaller firm, and its product, have been vetted to some degree, thus raising interest. Once one large prime contractor becomes aware of the small firm, then the reputation travels and others are likely to hear about it, its product, and area of expertise.

R&D-intensive companies were also said to have more opportunities to get to know each others' products through the SBIR process. For example, DARPA has previously hosted conferences for specific areas of technology, at which the prime contractors and smaller contractors displayed their work. While those conferences are not related to SBIR, they provide a venue through which firms interact, and a previous SBIR award will help set firms apart.

NanoSonic has, until now, not pursued venture-capital (VC) backing for any of its research projects. It was felt that venture capital is not an appropriate venue for research funding because VC firms often have unrealistic milestones for project development. They try to streamline companies by exercising extensive control (taking equity). When milestones are missed, repercussions can include losing additional ownership to the VC firm which, in extreme cases, assumes full control over the company and can break it up or change its focus. On the other hand, it was indicated that VC firms are more acceptable for product development, especially during the commercialization phase. They have the resources necessary to spread product information to the marketplace, and might have commercial connections of which the research firm is unaware or unable to utilize.

NanoSonic has also not pursued financing through bank loans. Banks are considered to know next to nothing about the business of small, R&D-intensive companies. The existence of an SBIR award would mean little in the consideration of financing opportunities.

In addition to providing NanoSonic with funding, SBIR awards have helped to build the company's technology base. They have allowed the company to pursue a whole suite of research projects and have allowed it to bring in researchers with various kinds of expertise that offer new points of view and knowledge that can be used to the company's benefit. Naturally, it was felt that not all ideas that receive SBIR funding turn out to be great. Some work perfectly and some don't work that well, with the majority falling somewhere in the middle, thus, forming a normal distribution of success.

EXAMPLE INNOVATION FROM NANOSONIC

An important example of innovation success stemming from the SBIR process for NanoSonic has been the so-called modified self-assembly process. Modified self-assembly is a process that allows the formation of multiple, ultrauniform, nanometer-thick layers of various materials into functional thin films. An early Air Force SBIR award in 1998 helped NanoSonic first develop the modified self-assembly process and apply it to the production of a special, transparent, electrically-conductive coating. This coating was composed of nanoparticles and could be applied to a wide variety of surfaces. Building on the patented process resulting from the initial SBIR grant, NanoSonic has managed to increase the volume of the material and the manufacturing speed of the modified self-assembly process.

A more recent SBIR award from DoE has allowed the company to apply this now-mature technology towards the creation of a product called Metal Rubber™. This product, formed from many layers of nanoparticles, combines the electrical conductivity of metals with a high level of plasticity, a reportedly unique combination. Additionally, the product is virtually optically transparent. Metal Rubber™ has applications in strengthening flexible electrical connections, such as those that power laptop screens and connect the flip panels of mobile phones.

It may also be used in areas such as electromagnetic shielding, numerous biomedical applications, and in measuring large mechanical deformations. This product, for which the company is very upbeat, is currently commercially available, through NanoSonic, for research and development applications in end-user processes only, or, by contract, for other development purposes.

Lockheed Martin has signed an alliance agreement with the firm in order to study the potential applications of nanocomposites in commercial and defense aerospace applications. Additionally, NanoSonic is associated with companies such as Motorola, General Dynamics, Dow Corning, Honeywell, IBM, and Northrop Grumman through partnerships, programs, contracts, or other agreements.

The firm has nine patents that have stemmed from the original, SBIR-sponsored, modified self-assembly process. It also has numerous other patents in other areas of nanoscale research. The modified self-assembly patents are exclusively licensed from Virginia Tech Intellectual Property, Inc. NanoSonic has separately developed its own intellectual property to allow for process, material, and product commercialization of outputs stemming from the modified self-assembly process.

VIEWS ON THE SBIR PROCESS

As stated earlier, NanoSonic was first exposed to the SBIR process very early in its history through an Air Force solicitation. According to the interviewees, the SBIR program is so well knows that it is difficult for any lab that performs technical research to miss it. In relation to the application process, NanoSonic does not feel that geography is important and it did not perceive important differences among the various agency application processes. The firm directs applications to the agency that makes the most sense for each specific research/innovation idea.

NanoSonic indicated that the costs associated with applying for SBIR funding are easily handled, as the learning process involved in putting together a strong application can actually be very beneficial, sometimes more than any eventual reward. As the company gathers information for completion of the proposal, it will contact a wide variety of industry representatives. The application process thus becomes a networking vehicle. It was indicated that well-prepared R&D firms can generate extensive intangible benefits from this networking, as potential research partners and future product marketers become more familiar with a specific innovation.

Overall, NanoSonic appeared quite satisfied with the SBIR application process. If anything, it was considered that providing instructions and incentives to small and large companies to facilitate pre-proposal networking would be very beneficial to all sides. NanoSonic's management perceives the SBIR program as a vehicle for getting small and large firms to cooperate, especially through the commercialization requirements attached to the grants. Agencies with an emphasis on commercialization should make every effort to strengthen such links.

No changes to the basic SBIR process were recommended. Topic specification, though it may vary in specification level across agency and program, was considered appropriate for each of the agency missions. The selection process is perceived as fair, with only a few cases where grantees seem to have had an early lock on the award. Feedback for rejected applications has tended to be straightforward and helpful. For innovations earning grants, the funding usually arrived in a timely manner, especially with the use of online filing. The interviewees encouraged all agencies to move to purely online filing, as this is a much more efficient method of grant application. Faced with the choice of fewer, larger grants or more, smaller grants, the interviewees thought current funding ratios appropriate: Moving to a smaller number of awards with larger funding was considered undesirable. The funds for testing feasibility in Phase I and testing production in Phase II were thought to be at more or less appropriate level.

All in all, NanoSonic sees the SBIR process as a very important framework that supports innovation in the United States. One of the reported strengths of the program is the requirement for feasibility studies. This requirement allows small companies to work alongside of, and network with, larger technology and production companies. Small companies benefit through the potential of aid from the large company resource base, and large companies benefit by being exposed to innovative technologies and products of smaller, nimbler firms. According to the interviewees, strengthening the program would mean further extending networking opportunities, for example, by granting small pre-proposal awards to test for product commercial potential. Conversely, NanoSonic sees few weaknesses in the overall program. Earlier problems regarding the timing of funding and application submission have been cleared up, principally due to a strong communication network. The firm did suggest that perhaps there could be more flexibility when it comes to spending grant monies on production. Otherwise, no significant changes were recommended.

NexTech Materials, Ltd.[24]

Nicholas S. Vonortas
Jeffrey Williams
The George Washington University

August 2005

THE COMPANY

NexTech Materials is located in Columbus, Ohio, and is focused on the development and manufacture of fuel cells and fuel-cell materials and components. Fuel cells use hydrogen or other fuels to produce electricity without generating pollutants. The company was founded in 1994 with only two employees and, having outgrown its 8,500-square-foot building in Worthington, recently moved into a 56,000-square-foot facility in Delaware County. It has about forty employees and sales of between $4 million and $5 million—both roughly double from two years ago.

Fuel-cell cars, although a potentially large market, are expected to be years away. Significant barriers remain, including how to best produce the needed hydrogen and make it available at service stations. NexTech works on various applications such as fuel cells for laptop computers, generators and other power sources. NexTech's first major commercial application of fuel-cell technology could be the home energy appliance in a few years; it is developing fuel-cell components for an appliance that would power and heat a home. The company expects demand for its products to increase as demonstration projects test fuel-cell technology. One project in Westerville will use fuel cells to generate electricity for 250 homes.

Fuel-cell materials and components are sold as either individual products, or combined and sold as a completed fuel cell. In 2000, Fuel Cell Materials was formed as a wholly owned subsidiary of NexTech in order to commercialize innovative ceramic materials technology and products developed by NexTech. The Fuel Cell Materials unit caters to other companies in the fuel cell industry through the leveraging of internal knowledge and production know-how. NexTech offers a range of partnership opportunities for other firms that operate in the fuel-

---

[24]This case is based on primary material collected by Nicholas Vonortas and Jeffrey Williams during a telephone interview with the President of NexTech Materials, Ltd., Mr. William Dawson. It is also based on preliminary research on the company carried out by the authors. We are indebted to Mr. Dawson for his willingness to participate and generosity in offering both a wealth of information to cover the various aspects of the study and his broad experience with the SBIR program and with high technology development in the context of small business. All opinions in the document are solely those of Mr. Dawson. The authors are responsible for remaining mistakes and misconceptions.

cell industry. Partnerships can include the provision of material components by NexTech as well as cooperative R&D based out of NexTech's facilities. Currently, the firm's projects include solid oxide fuel cell and oxygen generation components, fuel-cell processes and materials development, catalysts for fuel processing, sensors for fuel cells and fuel processing, and lead-free ceramics development.

NexTech has received numerous Phase I awards and five Phase II awards. It has never applied for Phase IIB but has applied unsuccessfully for Fast Track consideration from the Department of Defense. NexTech has successfully commercialized the results of a number of SBIR-funded projects, mostly with private-sector clients but also with federal agencies.

SBIR AND THE FIRM

The SBIR program has had a profound impact on the founding and subsequent success of NexTech. In 1994, Mr. Dawson started the company with his own money and funding from a Phase I SBIR contract with the U.S. Navy.[25] Without the SBIR funding, he argued, the firm would have probably never been incorporated, much less turned into a profitable enterprise. NexTech has maintained a close relationship with the SBIR program ever since, and the grants have been instrumental in seeding technology development and introducing the firm to commercial activities. Grant funding has made up between 6-25 percent of NexTech's revenue every year. That share is generally declining as commercializable innovations reach maturity. Mr. Dawson, however, indicated that SBIR awards may again account for an increased fraction of revenue in the future because fuel-cell technology makes up an emerging market. It is also currently difficult to accurately predict market opportunities in that area.

SBIR awards have assisted the company in attracting other forms of financing, such as funding opportunities from the state of Ohio's Third Frontier Fuel Cell Program and the Commerce Department's ATP program. The SBIR program has also been instrumental in introducing the firm to federal agencies, through Broad Area Announcements, as well as to other private firms. Contracts have been forthcoming from both of those interactions. Primarily, however, it has been private companies with which NexTech has developed the strongest commercial ties.

The SBIR program has contributed significantly to enhancing NexTech's technological base. Now that NexTech is more established and has a more focused portfolio, it is able to channel the SBIR funding into the development of new products in the firm's specialty base, as well as products derived from some of NexTech's existing innovations.

---

[25]Prior to founding the firm, Mr. Dawson worked as a project manager at the Battelle Memorial Institute. Since the inception of NexTech, he has earned three patents and authored numerous papers related to fuel cell technology. He also sits on the board of the U.S. Fuel Cell Council, the Ohio Fuel Cell Coalition, and the Edison Materials Technology Center.

EXAMPLE OF AN SBIR-DERIVED INNOVATION

The product selected for discussion was NexTech's low temperature solid oxide fuel cell. Improvement on the properties and performance of solid oxide fuel cells (SOFC) has been one of the main goals of the company since its inception. SOFC's directly convert chemical energy into electrical energy. Typically, a SOFC generates electricity by passing charged oxygen ions across a membrane. During the process, the oxygen ions are stripped of some electrical energy, which is then collected and transmitted as electricity. In this way, fuel cells function in a manner similar to that of regular, dry-cell batteries. However, fuel cells can be replenished when their chemical source of charged oxygen is depleted, thus allowing for a longer lifespan than traditional batteries.[26]

SOFC's are argued to be the most desirable form of fuel cell for a variety of reasons. First, SOFC's are more compact than other fuel cells because they use a solid electrolyte membrane, which negates the need for the pumps that are used in other cells to circulate the liquid membrane. Second, SOFC's are relatively highly efficient, transforming around 50 percent of chemical energy into electricity. Third, both carbon monoxide and hydrogen can be used as fuel sources. Being able to utilize more than one chemical as a fuel source ensures that the fuel supply for these cells is relatively inexpensive and plentiful.[27]

One of the drawbacks to SOFCs until now has been that they operate at high temperatures (at least 650° Celsius). At these temperatures, fuel cells require several minutes of start-up before being able to generate power, thus eliminating the possibility of use in instant-need electrical configurations, such as automobiles. NexTech, however, has created a SOFC that runs at comparatively lower temperatures, allowing them to be used in a wider variety of applications. NexTech SOFC's are currently used as power sources not only for stationary users, such as homes, hospitals, and other buildings, but also as continuously-running, auxiliary power sources for heavy trucks, as well as in a number of military applications.

The firm's fuel cells confer a number of benefits to their users. To begin with, given that the fuel-cell market is still relatively new, customers gain access to the new performance capabilities accompanying innovative products. NexTech's SOFC's also provide increased operating lifetimes and power output per unit weight over previous fuel cells. Many customers looking to use fuel-cell technology will be replacing traditional, entrenched forms of power generation. These fuel cells need then to not only match existing forms of power output but to appeal to users on different levels. In this case, fuel cells provide environmentally-friendly results, as they are nonpolluting, have zero negative emissions, and aid in the reduction of greenhouse gases.

---

[26]Azon.com Web site. Accessed at *<http://www.azom.com/details.asp?ArticleID=919&head=-Solid%2BOxide%2BFuel%2BCells>*.

[27]Ben Wiens Energy Science. Web site. *<http://www.benwiens.com/energy4.html>*.

NexTech is currently selling the materials used to build SOFCs on the global commercial market. For fuel-cell parts, the firm has approximately 170 customers in over thirty different countries. Complete SOFCs are also for sale globally. However, NexTech had only been selling complete cells on the open market for a few months at the time of the interview. Consequently, the market only covers Australia, Japan, Taiwan, and parts of Europe and North America. In addition to the commercial market, NexTech also sells fuel cells and fuel-cell components to the Department of Defense and to NASA.

Not only does NexTech develop and market their own products, they also offer commercial development services to other companies under contract. Other firms with their own fuel cell materials and products will approach NexTech for materials and development assistance that will improve those fuel-cell products and help them better penetrate the commercial marketplace. As such, NexTech can be considered a complete manufacturing and development concern built around fuel-cell technology.

NexTech's strategy for the commercialization of their low-temperature solid oxide fuel cells has been to remain a free agent. The firm supplies products and services to companies of all sizes, and seeks to supply the entire marketplace rather than concentrating on a small number of buyers. Under the assumption that, in the future, demand for fuel-cell products and services will increase, NexTech is also seeking strategic alliances through which manufacturing output can be increased. To date, the firm has not licensed its technology to other companies, but may consider doing so if the market grows beyond the point at which internal manufacturing capabilities can keep pace with demand. NexTech sees much more interest today for fuel-cell technology than in 1994 when the firm was founded, as fuel cells were a new innovation at that point. And while the market is not completely mature, current trends seem to support continued market growth, including steady annual increases in investment in fuel-cell development.

The firm sees a future fuel-cell market that will eventually grow to the size of multiple billions of dollars. That market, however, will not emerge at the same time for all applications of fuel cells. Areas in which NexTech is already active, such as heavy trucks and some military applications, are expected to reach maturity within five to ten years, while other areas, such as U.S. automotive and home markets, are still a long way from maturity. Parts of foreign fuel cell markets tend to be more mature than they are in the United States. For example, Germany, Switzerland, and Japan are already using fuel cells in residential power generation, an area that has been difficult to commercially penetrate in the United States. NexTech indicated that this is due to differences in the way marketplaces are structured, and that some sections of the U.S. market, such as fuel cells in heavy trucks, are probably more open than their foreign counterparts.

NexTech has published scientific papers related to the SOFC innovation. Frequently, NexTech publications occur as part of conferences in which the firm participates, rather than directly in peer-reviewed journals. The company is careful not to give away too much know-how, and focuses on publishing research results instead. Interestingly, Mr. Dawson indicated that it is more difficult for newly formed companies to preserve trade secrets in published materials. Consequently, when NexTech was first founded in 1994, the firm published very little in order to preserve its internally-generated tacit knowledge. In addition to published papers, two patents resulted from the SOFC innovation, and more are likely to come in the future.

VIEWS ON THE SBIR PROCESS

Before founding NexTech, Mr. Dawson was employed by the Battelle research consortium of Columbus, Ohio, a group that is active in support of entrepreneurial firms. It was there that he first heard of the SBIR grant process. Additionally, the State of Ohio actively promotes the SBIR program through conferences, as well as through email alerts distributed to innovation firms throughout the state.

While the SBIR program may be intended to have uniform functionality, NexTech does see differences among the granting agencies in a number of different areas. First, the firm generally applies for SBIR grants to the agency with which the specific technology has the best fit, as each agency has an individual technology focus. As has been also noted in other interviews, NSF tends to look for more basic research, DoE prefers a mix of basic and applied, and DoD looks for more applied technology. In terms of the SBIR applications, the only appreciable, difference according to the interviewee, is that some agencies, such as NSF, look for more formal commercialization plans than do others. One important area in which differences do occur is in the evaluation periods between payments. While switching to electronic filing and payments has sped up the overall process, some agencies do take longer than others in making payments. In NexTech's experience, the DoD is structured so that each individual SBIR-granting unit has its own payment process, which can cause delays from time to time. On NexTech's first SBIR grant, for example, Phase I was awarded in October of 1994, but payment for Phase I was not received until March of 1995. DoE and NSF, on the other hand, tend to be very responsive in payment distribution.

Another interesting issue brought up by Mr. Dawson also echoes comments in previous interviews. NexTech would prefer to see some sort of formal SBIR structure in place that helps prospective applicants before the agencies announce their topics. Having more information on the upcoming topic and on the individuals who will be reviewing applications reportedly results in much more successfully tailored proposals. In this interview, it was highlighted that being able to talk to people within the granting agency regarding the topic really helps the firm tailor their application and come forward with the most appropriate innovation. Given the significant proposal preparation costs, it was argued, it would be helpful to have an early indication of whether the granting agency would be interested at all in the innovation. The above concern is driven by the fact that, once a topic has been announced, firms are not permitted to communicate with program managers. Another possible change would be to increase the frequency of the rounds of topic specification. Especially in the DoD, it would be helpful if firms had more opportunities to respond to topic announcements during each year.

Regarding topic specification, NexTech recognizes that different agencies have different missions and must use topic specifications that are most appropriate for their needs. However, in the opinion of the interviewee, it is not enough for a firm to just respond to the topic announcement. Without having a working relationship with a program manager inside of the granting agency, there is little perceived chance that a firm will earn an SBIR award. This ties into NexTech's view on the overall fairness of the SBIR process. For the most part, the company sees it as very fair, but it considers that there is room for improvement.

One drawback of the process is when a firm's application is assigned to a program manager who does not have a lot of clout. In this case, equally effective innovations may suffer, as the more influential program managers are able to get their firms' innovations pushed through at the expense of other firms. Occasionally, there also might be cases in which application reviewers are specifically selected because they are known to typically give high or low marks. One final comment on fairness, which has been also heard in other interviews, is that some topics are so specific as to be only possibly filled by a single firm. While this is overall an undesirable action, NexTech did concede that defense agencies, in particular, need to find a way around budget restrictions and sponsor specific technologies that they cannot produce in-house.

Overall, NexTech sees SBIR as a very effective, very important program. However, some changes were suggested. While NexTech does not have experience with Phase IIB programs, they have applied once for a Fast Track award. A major roadblock was that the agency required the innovator to show proof that an investor had deposited third-party funding in the innovating firm's bank account before Fast Track could proceed. Letters of commitment are insufficient, and without the deposited funds, Fast Track contracts are canceled. Unfortunately for NexTech, investors with whom the firm has spoken are unwilling to deposit funds without a guarantee from the granting agency, thus making it difficult for the firm to participate in Fast Track programs. Combined with the lack of contract mechanisms to bridge Phase II and Phase III development, a number of innovations thus stand to remain unrealized.

Another suggested change was in the award amounts. Over the last ten years, SBIR grants have not increased in dollar value, making it harder for the money to be effective. If the funding were adjusted for inflation, that would go a long way towards eliminating the disparity. The relationship between Phase I and Phase II awards was also mentioned as an area that could use improvement. Specifically, the process needs to be more transparent in order to allow a firm to better understand its chances of earning a Phase II award after completing Phase I research. For NexTech, it seems that, sometimes, a great Phase I does not make it to Phase II, while a mediocre Phase I will earn Phase II funding.

Building on that last point, Mr. Dawson described the strategy of basing a company's revenues on SBIR awards as a bad business model due to its unpredictable nature. In a commercial contract, the customer and supplier have a clear indication that the relationship will involve an exchange of goods and services, allowing for a relatively accurate picture of income flow. With SBIR awards, it is more difficult to predict income flow, especially when the relationship between Phase I and Phase II is not transparent. Just as have many of the other interviewed firms, NexTech management does not support a business existing solely to earn SBIR grants. Not only does that operation plan suffer from the bad business model described above, but those firms also tend to disregard commercialization, which is the overall goal of the SBIR program.

Mr. Dawson sees the SBIR program today as being very different from 10 years back. No longer is it purely a policy system through which innovations are introduced into the marketplace. Instead, different agencies use the grant process to achieve different goals. NSF, for example, seeks to expand basic scientific knowledge with a strong emphasis on commercialization. DoD relies on these grants to get internal innovation projects going. The changing scope of military programs demands a large investment in innovative technology, but the DoD budget cannot support all of the innovation that the agency wants. Therefore, the SBIR program becomes a sort of contract vehicle through which specific firms and specific innovations are developed.

This last observation is related to a point that has been repeated in a number of the interviews: Having a personal relationship with the procurement side of an agency is essential to earning an SBIR award. NexTech employees dealing with the SBIR program are told that three things are required for a proposal to be successful: A great innovation idea, the innovation must to be important to the granting agency, and the program manager reviewing the proposal needs to have clout within the agency and be willing to push it through to contracts.

At the end, Mr. Dawson strongly reiterated his position that the SBIR program has been a boon for the nation's innovation industry. NexTech's history, he thought, is part of the proof.

Princeton Polymer Laboratories, Inc.[28]

Nicholas S. Vonortas
Jeffrey Williams
The George Washington University

July 2005

THE COMPANY

Princeton Polymer Laboratories, Inc (PPL) is primarily a contract research and development company based in Union, New Jersey. The firm was founded in 1969, well before the SBIR program came into being. Since the establishment of SBIR, PPL has earned five Phase I awards, from the Air Force and Department of Energy, and one Phase II award from the DoE. PPL has not fully commercialized any of its innovations. The rights to the Phase II innovation, a biopolymer, were purchased by Dupont-Conagra, but were subsequently returned to PPL when the buyer underwent a restructuring. PPL is 50 percent female-owned, and currently employs a mix of five full-time and contract researchers. Aside from R&D, firm employees frequently serve as expert witnesses in legal cases involving biotech issues.

SBIR AND THE FIRM

As the firm was founded in 1969, the SBIR program played no part in its establishment. And with only five Phase I awards and one Phase II award since the inception of SBIR, the program has played only a modest role in directly supporting the company financially. PPL's financing is currently derived entirely from the private sector. While the funding dollars themselves from SBIR have not made a large impact on the firm, the other benefits of the program have been more keenly felt. Most importantly for PPL, an SBIR award (to be discussed in the following section) allowed the firm to expand its technology base to include the biotech industry, which now comprises the bulk of its R&D activities. PPL was also able to secure funding, in the form of bridge grants, from the State of New Jersey at least partially due to the enhanced reputation associated with SBIR awardees.

---

[28]This case is based on primary material collected by Nicholas Vonortas and Jeffrey Williams during an interview with the President of Princeton Polymer Laboratories, Inc., Dr. Peter Wachtel. It is also based on preliminary research on the company carried out by the authors. We are indebted to Dr. Wachtel for his willingness to participate and generosity in offering both a wealth of information to cover the various aspects of the study and his broad experience with the SBIR program and with high technology development in the context of small business. All opinions in the document are solely those of Dr. Wachtel. The authors are responsible for remaining mistakes and misconceptions.

APPENDIX D *217*

EXAMPLE OF AN SBIR-DERIVED INNOVATION

Dr. Wachtel elected to discuss a biopolymer associated with the single Phase II award granted by the DoE. As background, a biopolymer is a biological, or biologically derived, synthetic polymer. In this case, the biopolymer is called chitosan, a naturally occurring material derived from the chitin of certain types of sea shells. Chitosan is used as a bulking agent in a number of commercial products, such as face creams, puddings, and diet supplements. It may also be used to aid in the extraction of heavy metals from waste-water, as it binds to the metals and causes them to clump and sink through the solution. There are also uses in the remediation of nuclear waste, as chitosan performs the same thickening action on uranium and plutonium that it does on other metals. The major drawbacks to the biological chitosan are that its production requires a lot of raw material—it takes around one ton of seashells to produce one pound of chitosan—and there are hazardous by-products associated with its production.

PPL sought to create a chitosan that performed as well as the existing type but did not have the toxic by-products. Through Phase I and Phase II funding from DoE, the firm created an insect-based form of chitosan that had no environmentally hazardous side effects. This product was initially licensed to the Dupont-Conagra cooperative concern. Unfortunately, however, Dupont-Conagra experienced financial difficulties soon thereafter and the new management decided to abandon the specific product development. The license was returned to PPL without having been commercialized.

Although promising, the form of chitosan developed with the Phase II award requires further development in order to be put into productive use. And like the seashell derived chitosan, it is still expensive to produce. Manufacturing can only be economically feasible when produced in industrial quantities at a large, dedicated facility and with round-the-clock staffing.[29] The firm is unable to produce the biopolymer in sufficient quantities in-house, and has not found any other outside enterprises willing to invest in its production. Currently, PPL has no specific plans for commercial development, though they would be interested in working with a corporate or government sponsor should the opportunity arise. Should such an arrangement occur, PPL would most likely sell the license to its chitosan, as it would be too large of a project for the firm to handle internally.

With respect to this product, PPL seems to be in a "Valley of Death" situation, where moving from Phase II prototype to product development and commercialization requires resources well beyond what the firm can muster (see annex of the case study). Were it able to secure sponsorship from an outside large firm or a government agency, PPL sees the potential market for its chitosan product as being very large, especially in the area of nuclear waste remediation. But it first requires a "patient" investor who can stay the course. Dr. Wachtel is

---

[29]Significant start-up costs.

not optimistic that this role would be filled by a large company under pressure for fast returns.

The chitosan process is a trade secret, and there are no plans for patenting. As a rule, PPL does not patent. It perceives process patents as particularly unnecessary to the scientific community in that they do not assist to retain intellectual property rights to processes. As a case in point, it was mentioned that one of PPL's previous owners spent large amounts of time and money on acquiring patents, but saw almost no return on those efforts. Along the same lines, PPL does not publish scientific papers out of concern of giving away proprietary information. PPL relies on its reputation and personal connections to attract clients, of which around 80 percent are repeat customers.

IMPRESSIONS OF THE SBIR PROCESS

Dr. Wachtel first became aware of the SBIR process in 1988, the same year in which he purchased PPL from his partners. PPL had a metal-polymer blend that had received some interest from the commercial sector. A personal connection at the Air Force then informed PPL of the SBIR program, and suggested that the firm submit the metal-polymer blend during the next round of SBIR funding, leading to PPL earning its first Phase I award. The firm has garnered awards from both DoD and DoE since 1988, and feels that there are some operational differences between the agencies. DoD, for example, rapidly issues an approval decision and funding for accepted submissions. DoE, on the other hand, is significantly slower to respond to inquiries or submissions.

PPL has not submitted any proposals for at least three years because the SBIR process is no longer seen as being cost effective: the amount of work necessary to submit proposals outweighs the resulting funding of those that are successful. Accordingly, Dr. Wachtel suggested two changes to the SBIR process. First, the approval process needs to be more transparent. It is difficult to tell what happens between proposal submission and the final decision by the granting agency. When asked about whether feedback helped in these instances, Dr. Wachtel indicated that the feedback is often very general, and mostly unusable. Even if the feedback were helpful, the SBIR process does not allow for a proposal to be reworked and resubmitted. Second, the submission topics are often too specific. Some are so precise that it gives the impression of the granting agency having a specific firm and technology in mind before announcing the submission round. However, he does acknowledge that some agencies need to be more specific than others as they have different needs and missions.

Overall, the impression is that SBIR is a good program because it is an important vehicle by which small firms are able to commercialize some innovations, though some improvements could be made to increase fairness.

PPL—ANNEX

Knowledge-intensive, innovative firms offer a return on investment that is skewed and highly uncertain, with risk characteristics and default probabilities that are hard to estimate. The likely existence of substantial informational asymmetries between such companies and investors make it difficult to come up with a mutually agreeable financing contract, since entrepreneurs may possess more information about the nature and characteristics of their products and processes than potential financiers. In addition, the intangible nature of innovative activities makes the assessment of their monetary values difficult before they become commercially successful and offers little salvage value in the event of failure. Regarding the firms, smaller companies tend to have limited market power, a lack of management skills, a higher share of intangible assets, an absence of adequate accounting track records and few assets, if any. The assessment can therefore be made that the more knowledge-intensive the firm, and the smaller its size, the harder it will be for it to gain access to capital.

This challenge of successfully moving from achieving a scientific breakthrough to creating a market-ready prototype is often referred to as the "Valley of Death." On one side of this valley stand the scientists and technologists, the innovators undertaking the research and development work; prior to reaching the "valley," they were funded through corporate or government research funds or—more rarely—from personal sources. On the other side stand innovation managers and investors, experts in financing and management of business enterprises; they possess development funds and expertise for turning an idea into a market-ready prototype supported by a validated business case.

Crossing the "Valley of Death" involves bridging three fundamental and interrelated gaps:[30]

- A financing gap between research funds—typically received from personal assets, government agencies or corporate research—that support more basic research and the investment funds to turn the idea into a market-ready prototype. This gap is usually bridged by risk financing through equity or by government programs specifically constructed for this purpose.
- A research gap between the scientific or technical breakthrough and the basis for a commercial product. Often, more research is needed on functionality, affordability, and quality before an idea can develop into a product that can compete in the marketplace.
- An information and trust gap between the scientist/technologist and the investor, each with a different understanding of the innovation and with dis-

---

[30]Lewis Branscomb and Philip Auerswald, *Taking Technical Risks: How Innovators, Executives and Investors Manage High-Tech Risks*, Cambridge, MA: The MIT Press, 2001.

similar expectations of what it is to accomplish. The technologist knows what is technically feasible and what is novel in the proposed approach; the investor knows the process of bringing new products to market. The two must be able to communicate effectively and to trust each other fully.

Thunderhead Engineering[31]

Philip E. Auerswald
George Mason University

September 2006

OVERVIEW

Thunderhead Engineering is a simulation software company located in Manhattan, Kansas, two hours to the west of Kansas City. The company was founded by in 1998 by Daniel Swenson, a professor in the Department of Mechanical and Nuclear Engineering at Kansas State University (K-State), in partnership with Brian Hardeman, then a master's degree student in the same department. Thunderhead is located in university town and was founded by academics in order to realize the commercial potential of capabilities developed in the process of university-conducted research. Though working far from both the technology centers on the two coasts and the oil industry hub in Houston, Thunderhead has succeeded in utilizing awards from the SBIR program to develop a simulation software product with stable customer base of major oil companies in the U.S. and overseas. Specifically, the company has built its business on tailoring for corporate use highly sophisticated simulation software developed at that Earth Sciences Division of Lawrence Berkeley National Laboratory.[32] Using its expertise in developing intuitive graphical user interfaces (GUIs) for complex engineering software, it has developed two relatively new products related to fire modeling and building design that have also achieved international sales.

FIRM DEVELOPMENT

Resisting the "Brain Drain:"
Two Academics Create Opportunity Where They Live

As much as it is about a technology, the story of the development of Thunderhead Engineering is about a place: K-State and Manhattan, Kansas. Company founder Swenson recalls his move from Sandia National Laboratory to Manhattan KS in mid-1980s: "I came to K-State primarily for family. My family

---

[31] This case is based primarily on primary material collected by Philip Auerswald during an interview at Thunderhead Engineering on September 30, 2005, with Dan Swenson and Brian Hardeman. We are indebted Thunderhead Engineering, Inc. for their willingness to participate in the study. Research assistance by Kirsten Apple is gratefully acknowledged. Views expressed in this case study are those of the authors, not of the National Academy of Sciences.

[32] Notably, the TOUGH2 and TOUGHRREACT software packages.

is from Kansas and I was looking for a place to get closer to them." Having chosen to live in Kansas for reasons unrelated to his professional development, Swenson sought upon his arrival to build up a research activity.

Brian Hardeman comments on the "brain drain" affecting Manhattan: "I used to say 'my wife and I are alone in this town because there is no one between the age of 22 and 40 because everyone graduates and they go get jobs elsewhere because there are not a lot of jobs here. There are not a lot of innovative high-tech companies. There are some manufacturing and services.' " For a young engineer a commitment to staying in Manhattan, Kansas—where Hardeman's wife is an elected local official—meant the need to get creative. Quips Hardeman: "I probably would not have a job if I did not have this company." Swenson is quick to qualify the comment, emphasizing that only the commitment to Manhattan, Kansas, has narrowed the range of Hardeman's career options. "Brian is someone that could have gone somewhere else and gotten a good job." Building Thunderhead in Manhattan, Kansas, was matter of choice, not necessity.

The partnership between Swenson and Hardeman began with a Department of Energy funded project on which the pair began work in 1996, with Swenson as the Principal Investigator and Hardeman the researcher. The objective of the project was to write software to model fluid flow and heat transfer in porous and fractured rocks. In the midst of that work, Swenson participated in a research conference in Japan. As a consequence of a presentation made during the trip, he received an offer to consult for a Japanese client. As a vehicle to perform this work, Swenson and Hardeman founded Thunderhead.

Using Open Source Code as the Basis for a Proprietary Software Package

Swenson and Hardeman began to consider the possibility of commercializing their software. "We did not have any clear plan," Hardeman recalls. The initial thought was "just to do something on the side in the evenings." As an initial step, they approached the K-State Technology Licensing Office.

The response they received was "eye opening," Hardeman recalls. The Technology Licensing Office was highly assertive of its claims on the software. "'If you want to use anything that has been developed at K-State, if you think there is even an inkling of money in it, we want licensing,'" Hardeman recounts as the essential message they received in their meeting. The university insisted that Thunderhead bear the cost of protecting the technology—a requirement that would have translated into a $10,000-$20,000 up-front payment. It was an attitude "that really shied us away from commercialization any of the work we had specifically done at K-State."

Soon thereafter, however, the two came across a Department of Energy SBIR solicitation with a topic they saw as "tailor-made" for their nascent company. The topic involved using software developed at Lawrence Livermore Lab—the leading competitor to the software that Swenson and Hardeman had developed at K-State—and building a graphical user interface to make it more accessible to commercial users. Hardeman recounts: "We had the company but we did not have a clear direction. Then this [SBIR solicitation] came along and provided us with perfect seed money for our company."

The motivation behind the solicitation was straightforward. Geothermal industry practitioners praised the technical quality of TOUGH and other simulation packages developed at the Lawrence Livermore, but at the same time they complained that the programs were excessively difficult to use—typically requiring a technician to train for three months before he or she could run the program and accurately interpret the results. Swenson and Hardeman appreciated the rigor and technical sophistication of the Lawrence Livermore code, but saw an opportunity in the relative ease of use of the program they had developed.

The Thunderhead Phase I application was successful. The company had cleared the most difficult hurdle, statistically, in participating in the SBIR program. The company's successful pitch in its Phase II application to DoE was that, although their simulation software did not have the potential to become "a huge money maker," the project leaders had demonstrated the capabilities needed to turn their work in Phase I into "a self-supporting continuing product that would be a great service to industry," with eventual applications in markets beyond geothermal.

From the standpoint of the development of the firm, the timing of the receipt of the first Phase II award was excellent. Swenson was due for a sabbatical year, having recently been granted tenure. Kansas State University covered half of Swenson's salary; the SBIR award covered the other half. The company had the resources to rent a modest office space, at a rate of $200/month, and to hire two K-State students to assist with programming. For two years, with little revenue beyond that from the SBIR award, the team focused on software development.

The award came to an end, but Swenson and Hardeman did not think that the product was ready to sell. "We continued to put in our own money and the 6 percent profit you get off a Phase II, to develop the product further. We worked with our students for about another six months." Finally, after nearly three years of effort, the team had a product that they could show to potential customers.

Once a product was ready, identifying potential customers was not difficult. "We knew everyone in geothermal," Swenson recalls. The greater challenge was arriving at a price for the product. "With any software product, pricing is kind of like throwing a dart at the wall." Without a clear point of reference, Thunderhead simply sought a price that they felt was fair to both them and their clients. Two dozen corporations signed on to annual agreements. The resulting income was modest from the standpoint of an SBIR Phase II award-recipient firm. From a pure commercialization standpoint, Swenson and Hardeman concede that the outcome would not have qualified the firm as a success if judged by the metrics used by the National Science Foundation's SBIR program—the funding agency for the company's subsequent awards in the program, noted for its particular focus on significant commercial outcomes.[33] However, from the Department of Energy standpoint, Thunderhead was an arguable success along other dimensions—in particular, making use of the outputs of research at a National Laboratory in support of agency mission.

Basing their marketing on a tightly knit group of contacts had the disadvantage that once that initial list of contacts was exhausted, the company struggled to reach additional potential customers. After 24 months during which sales reached a plateau, Thunderhead was able to "jump to the next level" after reaching an exclusive marketing arrangement with RockWare, Inc., a software distributor located in Golden, Colorado. Despite giving up 40 percent of every sale to the marketer, the company has realized a modest growth in revenue, achieving significantly greater reach with their product with substantially reduced effort.

Further validation of the value to industry of the company's software came in 2004 when Thunderhead reached an agreement with geothermal engineering teams at Shell, Exxon-Mobil, and Japan Power to jointly fund $45,000 of further development of the software—in part to take advantage of continued development of the underlying code by the Lawrence Berkeley Laboratory. Along similar lines, the value to agency mission was affirmed in 2004-2005 when the company received a $50,000 grant from the National Energy Technology Laboratory at the Department of Energy to add new capability to PetraSim to support methane hydrates. The company complemented these external sources of funds with investment of some of its internal resources to develop modules to extend the functionality of the core program.

Making the Transition from a Self-sustaining Product to a Self-sustaining Business

Even before their core product was utilized among geothermal engineers in the oil industry, Swenson and Hardeman were seeking the next challenge. "We were always looking for other opportunities because we knew [ours] was a niche product. . . . There was not a huge, great business case for this user interface for this geothermal software. People in industry were yearning for it. But there were not thousands of them—there were dozens."

"During our DoE Phase II we started to look at other things. NIST put out an SBIR solicitation for an interface for a fire modeling software," recalls Hardeman. The technical area was new to the team, but the match to their core capabilities was obvious. "It was a user interface again around core code software. So we responded."

This 2002 Phase I application was not successful. However, the signal back

---

[33]Hardeman elaborates: "NSF would not have funded [our first software development project] with out a larger business case. They are focused on a higher return business plan. They are really operating like venture capitalists." Accordingly, the pair notes that NSF SBIR topics are much more broadly defined than those in the DoE SBIR solicitation.

to them concerning their application was not clear, as NIST made only five awards for more than forty solicitations that year. After taking to the program manager they learned that, although the solicitation had appeared, the internal interest level for the topic was very low. The lack of commitment to the topic was frustrating. "I don't know how the topic got in there," Hardeman states. "They really did not intend to give an award."

Despite this frustration, the process of submitting the SBIR proposal did yield some benefits. Foremost among these was the set of contacts within the fire modeling industry that the company had made. When the proposal to NIST did not lead to an award, the group that Thunderhead had brought together decided to seek alternative sources of support for the project. Partnering with, Rolf Jensen & Associates, a fire engineering company, Thunderhead resubmitted the proposal to the National Science Foundation's SBIR program for a project in fire simulation. The company was on deck to enter a market an order of magnitude larger than that for their geothermal software. At that point, Swenson, states "we believed we could have a self-sustaining business not just a self-sustaining product."

The Innovation Element

While Thunderhead began with the objective of realizing the commercial value of federally funded research in geothermal simulation software, its continual development of a core product and entry into new markets has required it to innovate new approaches to modeling, simulation, and interface development. The resulting software "is not just an interface. It is quite applied." Have entered the fire modeling market, Thunderhead is now, according to Swenson, "starting to see the possibility of putting together a suit of building design fire protection tools." Thunderhead is now developing a companion product to model emergency egress from buildings. This couples the egress simulation to the fire model results, including blocked egress paths due to the fire.

A Tangible but Difficult to Measure SBIR Outcome: The Contribution to Community

Having made the commitment to Manhattan, Kansas, over a decade ago, Thunderhead's founder is now gratified to see that the company is beginning to function as a model to others in the community. "There is something that has really changed," Swenson reflects. After we received second Phase II, we began to earn reputation at K-state of being a legitimate company. This was partly a consequence of our interactions with our NSF program manger, who really wanted us to show that we could commercialize. So now I have faculty members coming to them asked would you go in with us and write a proposal—SBIR, STTR or another." The team has declined most of these offers, wanting to maintain a clear business direction instead of just being "the grant writers for Manhattan."

Patience and a focus on specific core capabilities have resulted in consistent success and steady growth. "We go after things that we (believe in or) are interested in and try and make living at it rather than look for that great huge opportunity and go after that," Hardeman notes. The focus has resulted in a Phase I success rate of 50 percent—more than double the program-wide average. The sales and contracting work that have accrued to the company from its PetraSim software developed with the DoE Phase I and Phase II awards admittedly do not qualify it as "a huge phase III with venture capital," in Hardeman's words. Yet the more $195,000 of sales and $100,000 of contract work that Thunderhead have earned on PetaSim have been enough to seed it as one of the relatively rare viable small technology companies operating in its environment. With the company having earned 45 percent of its lifetime revenues in the last 24 months, its growth clearly has not slowed. The company has three full-time employees and three part-time, with an increased focus on marketing. Student employees at Thunderhead are often eager to stay with company after their graduation.

To maintain continued growth the company's founder made the difficult decision in September 2005 to transition operational control to Hardeman. "I have made my decision—I am going to go back to K-State," Swenson states. "I am going to be phasing out from daily operations at Thunderhead. I will still be an owner and a board member. But it costs a lot to pay me or match my salary at K-State. We have a limited amount in the company and when I look at it, it would be better for me to back off. I think this is a better way to make it go."

SUMMARY

Because of the support provided by the SBIR program, Thunderhead Engineering has developed software that is meeting market opportunities. Both commercial products, PetraSim for simulation of flow in porous media and PyroSim for modeling of fires in buildings, were built around software developed at national laboratories, Lawrence Berkeley National Laboratory and the National Institute of Standards and Technology. This represents a leveraging of previous federal R&D investments to provide service a much broader set of beneficiaries than would otherwise be possible.

Thunderhead Engineering is now on track to be self-supporting through sales. Sales of PyroSim are steadily increasing and the new emergency egress software will integrate with the existing product. There is every reason to believe that Thunderhead Engineering will continue to grow, thanks to the initial SBIR support.

Appendix E

Bibliography

Acs, Z., and D. Audretsch. 1991. *Innovation and Small Firms.* Cambridge, MA: MIT Press.
Advanced Technology Program. 2001. *Performance of 50 Completed ATP Projects, Status Report 2.* National Institute of Standards and Technology Special Publication 950-2. Washington, DC: Advanced Technology Program/National Institute of Standards and Technology/U.S. Department of Commerce.
Alic, John A., Lewis Branscomb, Harvey Brooks, Ashton B. Carter, and Gerald L. Epstein. 1992. *Beyond Spinoff: Military and Commercial Technologies in a Changing World.* Boston, MA: Harvard Business School Press.
American Association for the Advancement of Science. "R&D Funding Update on NSF in the FY2007." Available online at <http://www.aaas.org/spp/rd/nsf07hf1.pdf>.
Archibald, R., and D. Finifter. 2000. "Evaluation of the Department of Defense Small Business Innovation Research Program and the Fast Track Initiative: A Balanced Approach." In National Research Council. *The Small Business Innovation Research Program: An Assessment of the Department of Defense Fast Track Initiative.* Charles W. Wessner, ed. Washington, DC: National Academy Press.
Arrow, Kenneth. 1962. "Economic welfare and the allocation of resources for Invention." Pp. 609–625 in *The Rate and Direction of Inventive Activity: Economic and Social Factors.* Princeton, NJ: Princeton University Press.
Arrow, Kenneth. 1973. "The theory of discrimination." Pp. 3–31 in *Discrimination in Labor Market.* Orley Ashenfelter and Albert Rees, eds. Princeton, NJ: Princeton University Press.
Audretsch, David B. 1995. *Innovation and Industry Evolution.* Cambridge, MA: MIT Press.
Audretsch, David B., and Maryann P. Feldman. 1996. "R&D spillovers and the geography of innovation and production." *American Economic Review* 86(3):630–640.
Audretsch, David B., and Paula E. Stephan. 1996. "Company-scientist locational links: The case of biotechnology." *American Economic Review* 86(3):641–642.
Audretsch, D., and R. Thurik. 1999. *Innovation, Industry Evolution, and Employment.* Cambridge, MA: MIT Press.
Baker, Alan. No date. "Commercialization Support at NSF." Draft.
Barfield, C., and W. Schambra, eds. 1986. *The Politics of Industrial Policy.* Washington, DC: American Enterprise Institute for Public Policy Research.

Baron, Jonathan. 1998. "DoD SBIR/STTR Program Manager." Comments at the Methodology Workshop on the Assessment of Current SBIR Program Initiatives, Washington, DC, October.

Barry, C. B. 1994. "New directions in research on venture capital finance." *Financial Management* 23 (Autumn):3–15.

Bator, Francis. 1958. "The anatomy of market failure." *Quarterly Journal of Economics* 72:351–379.

Bingham, R. 1998. *Industrial Policy American Style: From Hamilton to HDTV*. New York: M.E. Sharpe.

Birch, D. 1981. "Who Creates Jobs." *The Public Interest* 65 (Fall):3–14.

Branscomb, Lewis M. 2000. *Managing Technical Risk: Understanding Private Sector Decision Making on Early Stage Technology Based Projects*. Washington, DC: Department of Commerce/National Institute of Standards and Technology.

Branscomb, Lewis M., and Philip E. Auerswald. 2001. *Taking Technical Risks: How Innovators, Managers, and Investors Manage Risk in High-Tech Innovations*, Cambridge, MA: MIT Press.

Branscomb, Lewis M., and J. Keller. 1998. *Investing in Innovation: Creating a Research and Innovation Policy*. Cambridge, MA: MIT Press.

Brav, A., and P. A. Gompers. 1997. "Myth or reality?: Long-run underperformance of initial public offerings; Evidence from venture capital and nonventure capital-backed IPOs." *Journal of Finance* 52:1791–1821.

Brodd, R. J. 2005. *Factors Affecting U.S. Production Decisions: Why Are There No Volume Lithium-Ion Battery Manufacturers in the United States?* ATP Working Paper No. 05-01, June 2005.

Brown, G., and J. Turner. 1999. "Reworking the Federal Role in Small Business Research." *Issues in Science and Technology* XV, no. 4 (Summer).

Bush, Vannevar. 1946. *Science—the Endless Frontier*. Republished in 1960 by U.S. National Science Foundation, Washington, DC.

Carden, S. D. and O. Darragh. 2004. "A Halo for Angel Investors." *The McKinsey Quarterly* 1.

Cassell, G. 2004. "Setting Realistic Expectations for Success." In National Research Council. *SBIR: Program Diversity and Assessment Challenges*. Charles W. Wessner, ed. Washington, DC: The National Academies Press.

Caves, Richard E. 1998. "Industrial organization and new findings on the turnover and mobility of firms." *Journal of Economic Literature* 36(4):1947–1982.

Clinton, William Jefferson. 1994. *Economic Report of the President*. Washington, DC: U.S. Government Printing Office.

Clinton, William Jefferson. 1994. *The State of Small Business*. Washington, DC: U.S. Government Printing Office.

Coburn, C., and D. Bergland. 1995. *Partnerships: A Compendium of State and Federal Cooperative Technology Programs*. Columbus, OH: Battelle.

Cochrane, J. 2005. "The Risk and Return of Venture Capital." *Journal of Financial Economics* 75(1): 3-52.

Cohen, L. R., and R. G. Noll. 1991. *The Technology Pork Barrel*. Washington, DC: The Brookings Institution.

Congressional Commission on the Advancement of Women and Minorities in Science, Engineering, and Technology Development. 2000. *Land of Plenty: Diversity as America's Competitive Edge in Science, Engineering and Technology*. Washington, DC: National Science Foundation/U.S. Government Printing Office.

Cooper, R.G. 2001. *Winning at New Products: Accelerating the process from idea to launch*. In Dawnbreaker, Inc. 2005. "The Phase III Challenge: Commercialization Assistance Programs 1990–2005." White paper. July 15.

Council of Economic Advisers. 1995. *Supporting Research and Development to Promote Economic Growth: The Federal Government's Role*. Washington, DC.

Cramer, Reid. 2000. "Patterns of Firm Participation in the Small Business Innovation Research Program in Southwestern and Mountain States." In National Research Council. 2000. *The Small Business Innovation Research Program: An Assessment of the Department of Defense Fast Track Initiative.* Charles W. Wessner, ed. Washington, DC: National Academy Press.

Davis, S. J., J. Haltiwanger, and S. Schuh. 1994. "Small Business and Job Creation: Dissecting the Myth and Reassessing the Facts," *Business Economics* 29(3):113–122.

Dawnbreaker, Inc. 2005. "The Phase III Challenge: Commercialization Assistance Programs 1990–2005." White paper. July 15.

Dertouzos. 1989. *Made in America: The MIT Commission on Industrial Productivity.* Cambridge, MA: MIT Press.

DoE Opportunity Forum. 2005. "Partnering and Investment Opportunities for the Future." Tysons Corner, VA. October 24–25.

Eckstein, O. 1984. *DRI Report on U.S. Manufacturing Industries.* New York: McGraw Hill.

Eisinger, P. K. 1988. *The Rise of the Entrepreneurial State: State and Local Economic Development Policy in the United State.* Madison, WI: University of Wisconsin Press.

Feldman, Maryann P. 1994a. "Knowledge complementarity and innovation." *Small Business Economics* 6(5):363–372.

Feldman, Maryann P. 1994b. *The Geography of Knowledge.* Boston, MA: Kluwer Academic.

Feldman, M. P., and M. R. Kelley. 2001. *Winning an Award from the Advanced Technology Program: Pursuing R&D Strategies in the Public Interest and Benefiting from a Halo Effect.* NISTIR 6577. Washington, DC: Advanced Technology Program/National Institute of Standards and Technology/U.S. Department of Commerce.

Fenn, G. W., N. Liang, and S. Prowse. 1995. *The Economics of the Private Equity Market.* Washington, DC: Board of Governors of the Federal Reserve System.

Flamm, K. 1988. *Creating the Computer.* Washington, DC: The Brookings Institution.

Freear, J., and W. E. Wetzel Jr. 1990. "Who bankrolls high-tech entrepreneurs?" *Journal of Business Venturing* 5:77–89.

Freeman, Chris, and Luc Soete. 1997. *The Economics of Industrial Innovation.* Cambridge, MA: MIT Press.

Galbraith, J. K. 1957. *The New Industrial State.* Boston: Houghton Mifflin.

Geroski, Paul A. 1995. "What do we know about entry?" *International Journal of Industrial Organization* 13(4):421–440.

Gompers, P. A., and J. Lerner. 1977. "Risk and Reward in Private Equity Investments: The Challenge of Performance Assessment." *Journal of Private Equity* 1(Winter):5-12.

Gompers, P. A. 1995. "Optimal investment, monitoring, and the staging of venture capital." *Journal of Finance* 50:1461–1489.

Gompers, P. A., and J. Lerner. 1996. "The use of covenants: An empirical analysis of venture partnership agreements." *Journal of Law and Economics* 39:463–498.

Gompers, P. A., and J. Lerner. 1998a. "What drives venture capital fund-raising?" Unpublished working paper. Harvard University.

Gompers, P. A., and J. Lerner. 1998b. "Capital formation and investment in venture markets: A report to the NBER and the Advanced Technology Program." Unpublished working paper. Harvard University.

Gompers, P. A., and J. Lerner. 1999. "An analysis of compensation in the U.S. venture capital partnership." *Journal of Financial Economics* 51(1):3–7.

Gompers, P. A., and J. Lerner. 1999. *The Venture Cycle.* Cambridge, MA: MIT Press.

Good, M. L. 1995. Prepared testimony before the Senate Commerce, Science, and Transportation Committee, Subcommittee on Science, Technology, and Space (photocopy, U.S. Department of Commerce).

Goodnight, J. 2003. Presentation at National Research Council Symposium. "The Small Business Innovation Research Program: Identifying Best Practice." Washington, DC May 28.

Graham, O. L. 1992. *Losing Time: The Industrial Policy Debate*. Cambridge, MA: Harvard University Press.

Greenwald, B. C., J. E. Stiglitz, and A. Weiss. 1984. "Information imperfections in the capital market and macroeconomic fluctuations." *American Economic Review Papers and Proceedings* 74:194–199.

Griliches, Z. 1990. *The Search for R&D Spillovers*. Cambridge, MA: Harvard University Press.

Hall, Bronwyn H. 1992. "Investment and research and development: Does the source of financing matter?" Working Paper No. 92–194, Department of Economics/University of California at Berkeley.

Hall, Bronwyn H. 1993. "Industrial research during the 1980s: Did the rate of return fall?" Brookings Papers: *Microeconomics* 2:289–343.

Hamberg, Dan. 1963. "Invention in the industrial research laboratory." *Journal of Political Economy* (April):95–115.

Hao, K. Y., and A. B. Jaffe. 1993. "Effect of liquidity on firms' R&D spending." *Economics of Innovation and New Technology* 2:275–282.

Hebert, Robert F., and Albert N. Link. 1989. "In search of the meaning of entrepreneurship." *Small Business Economics* 1(1):39–49.

Himmelberg, C. P., and B. C. Petersen. 1994. "R&D and internal finance: A panel study of small firms in high-tech industries." *Review of Economics and Statistics* 76:38–51.

Hubbard, R. G. 1998. "Capital-market imperfections and investment." *Journal of Economic Literature* 36:193–225.

Huntsman, B., and J. P. Hoban Jr. 1980. "Investment in new enterprise: Some empirical observations on risk, return, and market structure." *Financial Management* 9 (Summer):44–51.

Jaffe, A. B. 1996. "Economic Analysis of Research Spillovers: Implications for the Advanced Technology Program." Washington, DC: Advanced Technology Program/National Institute of Standards and Technology/U.S. Department of Commerce).

Jaffe, A. B. 1998a. "Economic Analysis of Research Spillovers: Implications for the Advanced Technology Program." Washington, DC: Advanced Technology Program/National Institute of Standards and Technology/U.S. Department of Commerce.

Jaffe, A. B. 1998b. "The importance of 'spillovers' in the policy mission of the Advanced Technology Program." *Journal of Technology Transfer* (Summer).

Jewkes, J., D. Sawers, and R. Stillerman. 1958. *The Sources of Invention*. New York: St. Martin's Press.

Johnson, M. 2004. "SBIR at the Department of Energy: Achievements, Opportunities, and Challenges." In National Research Council. *SBIR: Program Diversity and Assessment Challenges*. Charles W. Wessner, ed. Washington, D.C.: The National Academies Press, 2004.

Kauffman Foundation. About the Foundation. Available online at <*http://www.kauffman.org/foundation.cfm*>.

Kleinman, D. L. 1995. *Politics on the Endless Frontier: Postwar Research Policy in the United States*. Durham, NC: Duke University Press.

Kortum, Samuel, and Josh Lerner. 1998. "Does Venture Capital Spur Innovation?" NBER Working Paper No. 6846, National Bureau of Economic Research.

Krugman, P. 1990. *Rethinking International Trade*. Cambridge, MA: MIT Press.

Krugman, P. 1991. *Geography and Trade*. Cambridge, MA: MIT Press.

Langlois, Richard N., and Paul L. Robertson. 1996. "Stop Crying over Spilt Knowledge: A Critical Look at the Theory of Spillovers and Technical Change." Paper prepared for the MERIT Conference on Innovation, Evolution, and Technology. Maastricht, Netherlands, August 25–27.

Lebow, I. 1995. *Information Highways and Byways: From the Telegraph to the 21st Century*. New York: Institute of Electrical and Electronic Engineering.

Lerner, J. 1994. "The syndication of venture capital investments." *Financial Management* 23 (Autumn):16–27.

Lerner, J. 1995. "Venture capital and the oversight of private firms." *Journal of Finance* 50:301–318.

Lerner, J. 1996. "The government as venture capitalist: The long-run effects of the SBIR program." Working Paper No. 5753, National Bureau of Economic Research.

Lerner, J. 1998. "Angel financing and public policy: An overview." *Journal of Banking and Finance* 22(6–8):773–784.

Lerner, J. 1999a. "The government as venture capitalist: The long-run effects of the SBIR program." *Journal of Business* 72(3):285–297.

Lerner, J. 1999b. "Public venture capital: Rationales and evaluation." In *The SBIR Program: Challenges and Opportunities*. Washington, DC: National Academy Press.

Levy, D. M., and N. Terleckyk. 1983. "Effects of government R&D on private R&D investment and productivity: A macroeconomic analysis." *Bell Journal of Economics* 14:551–561.

Liles, P. 1977. *Sustaining the Venture Capital Firm*. Cambridge, MA: Management Analysis Center.

Link, Albert N. 1998. "Public/Private Partnerships as a Tool to Support Industrial R&D: Experiences in the United States." Paper prepared for the working group on Innovation and Technology Policy of the OECD Committee for Science and Technology Policy, Paris.

Link, Albert N., and John Rees. 1990. "Firm size, university based research and the returns to R&D." *Small Business Economics* 2(1):25–32.

Link, Albert N., and John T. Scott. 1998a. "Assessing the infrastructural needs of a technology-based service sector: A new approach to technology policy planning." *STI Review* 22:171–207.

Link, Albert N., and John T. Scott. 1998b. *Overcoming Market Failure: A Case Study of the ATP Focused Program on Technologies for the Integration of Manufacturing Applications (TIMA)*. Draft final report submitted to the Advanced Technology Program. Gaithersburg, MD: National Institute of Technology. October.

Link, Albert N., and John T. Scott. 1998c. *Public Accountability: Evaluating Technology-Based Institutions*. Norwell, MA: Kluwer Academic.

Link, A. N., and J. T. Scott. 2005. *Evaluating Public Research Institutions: The U.S. Advanced Technology Program's Intramural Research Initiative*. London: Routledge.

Longini, P. 2003. "Hot buttons for NSF SBIR Research Funds," Pittsburgh Technology Council, *TechyVent*. November 27.

Malone, T. 1995. *The Microprocessor: A Biography*. Hamburg, Germany: Springer Verlag/Telos.

Mansfield, E. 1985. "How Fast Does New Industrial Technology Leak Out?" *Journal of Industrial Economics* 34(2).

Mansfield, E. 1996. *Estimating Social and Private Returns from Innovations Based on the Advanced Technology Program: Problems and Opportunities*. Unpublished report.

Mansfield, E., J. Rapoport, A. Romeo, S. Wagner, and G. Beardsley. 1977. "Social and private rates of return from industrial innovations." *Quarterly Journal of Economics* 91:221–240.

Martin, Justin. 2002. "David Birch." *Fortune Small Business* (December 1).

McCraw, T. 1986. "Mercantilism and the Market: Antecedents of American Industrial Policy." In C. Barfield and W. Schambra, eds. *The Politics of Industrial Policy*. Washington, DC: American Enterprise Institute for Public Policy Research.

Mervis, Jeffrey D. 1996. "A $1 Billion 'Tax' on R&D Funds." *Science* 272:942–944.

Moore, D. 2004. "Turning Failure into Success." In National Research Council. *The Small Business Innovation Research Program: Program Diversity and Assessment Challenges*. Charles W. Wessner, ed. Washington, DC: The National Academies Press.

Mowery, D. 1998. "Collaborative R&D: how effective is it?" *Issues in Science and Technology* (Fall):37–44.

Mowery, D., ed. 1999. *U.S. Industry in 2000: Studies in Competitive Performance*. Washington, DC: National Academy Press.

Mowery, D., and N. Rosenberg. 1989. *Technology and the Pursuit of Economic Growth*. New York: Cambridge University Press.

Mowery, D., and N. Rosenberg. 1998. *Paths of Innovation: Technological Change in 20th Century America*. New York: Cambridge University Press.

Myers, S., R. L. Stern, and M. L. Rorke. 1983. *A Study of the Small Business Innovation Research Program*. Lake Forest, IL: Mohawk Research Corporation.

Myers, S. C., and N. Majluf. 1984. "Corporate financing and investment decisions when firms have information that investors do not have." *Journal of Financial Economics* 13:187–221.

National Research Council. 1986. *The Positive Sum Strategy: Harnessing Technology for Economic Growth*. Washington, DC: National Academy Press.

National Research Council. 1987. *Semiconductor Industry and the National Laboratories: Part of a National Strategy*. Washington, DC: National Academy Press.

National Research Council. 1991. *Mathematical Sciences, Technology, and Economic Competitiveness*. James G. Glimm, ed. Washington, DC: National Academy Press.

National Research Council. 1992. *The Government Role in Civilian Technology: Building a New Alliance*. Washington, DC: National Academy Press.

National Research Council. 1995. *Allocating Federal Funds for R&D*. Washington, DC: National Academy Press.

National Research Council. 1996. *Conflict and Cooperation in National Competition for High-Technology Industry*. Washington, DC: National Academy Press.

National Research Council. 1997. *Review of the Research Program of the Partnership for a New Generation of Vehicles: Third Report*. Washington, DC: National Academy Press.

National Research Council. 1999a. *The Advanced Technology Program: Challenges and Opportunities*. Charles W. Wessner, ed. Washington, DC: National Academy Press.

National Research Council. 1999b. *Funding a Revolution: Government Support for Computing Research*. Washington, DC: National Academy Press.

National Research Council. 1999c. *Industry-Laboratory Partnerships: A Review of the Sandia Science and Technology Park Initiative*. Charles W. Wessner, ed. Washington, DC: National Academy Press.

National Research Council. 1999d. *New Vistas in Transatlantic Science and Technology Cooperation*. Charles W. Wessner, ed. Washington, DC: National Academy Press.

National Research Council. 1999e. *The Small Business Innovation Research Program: Challenges and Opportunities*. Charles W. Wessner, ed. Washington, DC: National Academy Press.

National Research Council. 2000a. *The Small Business Innovation Research Program: A Review of the Department of Defense Fast Track Initiative*. Charles W. Wessner, ed. Washington, DC: National Academy Press.

National Research Council. 2000b. *U.S. Industry in 2000: Studies in Competitive Performance*. Washington, DC: National Academy Press.

National Research Council. 2001a. *The Advanced Technology Program: Assessing Outcomes*. Charles W. Wessner, ed. Washington, DC: National Academy Press.

National Research Council. 2001b. *Attracting Science and Mathematics Ph.Ds to Secondary School Education*. Washington, DC: National Academy Press.

National Research Council. 2001c. *Building a Workforce for the Information Economy*. Washington, DC: National Academy Press.

National Research Council. 2001d. *Capitalizing on New Needs and New Opportunities: Government-Industry Partnerships in Biotechnology and Information Technologies*. Charles W. Wessner, ed. Washington, DC: National Academy Press.

National Research Council. 2001e. *A Review of the New Initiatives at the NASA Ames Research Center*. Charles W. Wessner, ed. Washington, DC: National Academy Press.

National Research Council. 2001f. *Trends in Federal Support of Research and Graduate Education*. Washington, DC: National Academy Press.

National Research Council. 2002a. *Government-Industry Partnerships for the Development of New Technologies: Summary Report*. Charles W. Wessner, ed. Washington, DC: The National Academies Press.

National Research Council. 2002b. *Measuring and Sustaining the New Economy*. Dale W. Jorgenson and Charles W. Wessner, eds. Washington, DC: National Academy Press.

APPENDIX E   233

National Research Council. 2002c. *Partnerships for Solid-State Lighting*. Charles W. Wessner, ed. Washington, DC: The National Academies Press.
National Research Council. 2004a. *An Assessment of the Small Business Innovation Research Program: Project Methodology*. Washington, DC: The National Academies Press.
National Research Council. 2004b. Capitalizing on Science, Technology, and Innovation: An Assessment of the Small Business Innovation Research Program/Program Manager Survey. Completed by Dr. Joseph Hennessey.
National Research Council. 2004c. *Productivity and Cyclicality in Semiconductors: Trends, Implications, and Questions.*. Dale W. Jorgenson and Charles W. Wessner, eds. Washington, DC: The National Academies Press.
National Research Council. 2004d. *The Small Business Innovation Research Program: Program Diversity and Assessment Challenges*. Charles W. Wessner, ed. Washington, DC: The National Academies Press.
National Research Council. 2006a. *Beyond Bias and Barriers: Fulfilling the Potential of Women in Academic Science and Engineering*.
National Research Council. 2006b. *Deconstructing the Computer*. Dale W. Jorgenson and Charles W. Wessner, eds. Washington, DC: The National Academies Press.
National Research Council. 2006c. Capitalizing on Science, Technology, and Innovation: An Assessment of the Small Business Innovation Research Program/Firm Survey.
National Research Council. 2006d. Capitalizing on Science, Technology, and Innovation: An Assessment of the Small Business Innovation Research Program/Phase I Survey.
National Research Council. 2006e. Capitalizing on Science, Technology, and Innovation: An Assessment of the Small Business Innovation Research Program/Phase II Survey.
National Research Council. 2006f. *Software, Growth, and the Future of the U.S. Economy*. Dale W. Jorgenson and Charles W. Wessner, eds. Washington, DC: The National Academies Press.
National Research Council. 2006g. *The Telecommunications Challenge: Changing Technologies and Evolving Policies*. Dale W. Jorgenson and Charles W. Wessner, eds. Washington, DC: The National Academies Press.
National Research Council. 2007a. *Enhancing Productivity Growth in the Information Age: Measuring and Sustaining the New Economy*. Dale W. Jorgenson and Charles W. Wessner, eds. Washington, DC: The National Academies Press.
National Research Council. 2007b. *India's Changing Innovation System: Achievements, Challenges, and Opportunities for Cooperation*. Charles W. Wessner and Sujai J. Shivakumar, eds. Washington, DC: The National Academies Press.
National Research Council. 2007c. *Innovation Policies for the 21st Century*. Charles W. Wessner, ed. Washington, DC: The National Academies Press.
National Research Council. 2007d. *SBIR and the Phase III Challenge of Commercialization*. Charles W. Wessner, ed. Washington, DC: The National Academies Press.
National Science Foundation. Committee of Visitors Reports and Annual Updates. Available online at <http://www.nsf.gov/eng/general/cov/>.
National Science Foundation. Emerging Technologies. Available online at <http://www.nsf.gov/eng/sbir/eo.jsp>.
National Science Foundation. Guidance for Reviewers. Available online at <http://www.eng.nsf.gov/sbir/peer_review.htm>.
National Science Foundation. National Science Foundation at a Glance. Available online at <http://www.nsf.gov/about>.
National Science Foundation. National Science Foundation Manual 14, *NSF Conflicts of Interest and Standards of Ethical Conduct*. Available online at <http://www.eng.nsf.gov/sbir/COI_Form.doc>.
National Science Foundation. The Phase IIB Option. Available online at <http://www.nsf.gov/eng/sbir/phase_IIB.jsp#ELIGIBILITY>.

National Science Foundation. Proposal and Grant Manual. Available online at <http://www.inside.nsf.gov/pubs/2002/pam/pamdec02.6html>.
National Science Foundation. 2005. Synopsis of SBIR/STTR Program. Available online at <http://www.nsf.gov/funding/pgm_summ.jsp?Phase Ims_id=13371&org=DMII>.
National Science Foundation. 2006. "News items from the past year." Press Release. April 10.
National Science Foundation, Office of Industrial Innovation. 2006. "SBIR/STTR Phase II Grantee Conference, Book of Abstracts." Louisville, Kentucky. May 18–20, 2006.
National Science Foundation, Office of Industrial Innovation. Draft Strategic Plan, June 2, 2005.
National Science Foundation, Office of Legislative and Public Affairs. 2003. SBIR Success Story from News Tip. Web's "Best Meta-Search Engine," March 20.
National Science Foundation, Office of Legislative and Public Affairs. 2004. SBIR Success Story: GPRA Fiscal Year 2004 "Nugget." Retrospective Nugget–AuxiGro Crop Yield Enhancers.
Nelson, R. R. 1982. *Government and Technological Progress*. New York: Pergamon.
Nelson, R. R. 1986. "Institutions supporting technical advances in industry." *American Economic Review, Papers and Proceedings* 76(2):188.
Nelson, R. R., ed. 1993. *National Innovation System: A Comparative Study*. New York: Oxford University Press.
Office of Management and Budget. 1996. "Economic analysis of federal regulations under Executive Order 12866."
Office of the President. 1990. *U.S. Technology Policy*. Washington, DC: Executive Office of the President.
Organization for Economic Cooperation and Development. 1982. *Innovation in Small and Medium Firms*. Paris: Organization for Economic Cooperation and Development.
Organization for Economic Cooperation and Development. 1995. *Venture Capital in OECD Countries*. Paris: Organization for Economic Cooperation and Development.
Organization for Economic Cooperation and Development. 1997. *Small Business Job Creation and Growth: Facts, Obstacles, and Best Practices*. Paris: Organization for Economic Cooperation and Development.
Organization for Economic Cooperation and Development. 1998. *Technology, Productivity and Job Creation: Toward Best Policy Practice*. Paris: Organization for Economic Cooperation and Development.
Pacific Northwest National Laboratory. SBIR Alerting Service. Available online at <http://www.pnl.gov/edo/sbir>.
Perko, J. S., and F. Narin. 1997. "The Transfer of Public Science to Patented Technology: A Case Study in Agricultural Science." *Journal of Technology Transfer* 22(3):65–72.
Perret, G. 1989. *A Country Made by War: From the Revolution to Vietnam—The Story of America's Rise to Power*. New York: Random House.
Porter, M. 1998. "Clusters and Competition: New Agendas for Government and Institutions." In *On Competition*. Boston, MA: Harvard Business School Press.
Powell, J. W. 1999. *Business Planning and Progress of Small Firms Engaged in Technology Development through the Advanced Technology Program*. NISTIR 6375. National Institute of Standards and Technology/U.S. Department of Commerce.
Powell, Walter W., and Peter Brantley. 1992. "Competitive cooperation in biotechnology: Learning through networks?" In N. Nohria and R. G. Eccles, eds. *Networks and Organizations: Structure, Form and Action*. Boston, MA: Harvard Business School Press. Pp. 366–394.
Price Waterhouse. 1985. *Survey of small high-tech businesses shows Federal SBIR awards spurring job growth, commercial sales*. Washington, DC: Small Business High Technology Institute.
Roberts, Edward B. 1968. "Entrepreneurship and technology." *Research Management* (July):249–266.
Romer, P. 1990. "Endogenous technological change." *Journal of Political Economy* 98:71–102.
Rosa, R. and A. Dawson. 2006. "Gender and the Commercialization of University Science: Academic Founders of Spinout Companies." *Entrepreneurship & Regional Development* 18(4):341-366.

Rosenbloom, R., and Spencer, W. 1996. *Engines of Innovation: U.S. Industrial Research at the End of an Era*. Boston, MA: Harvard Business School Press.

Rubenstein, A. H. 1958. *Problems Financing New Research-Based Enterprises in New England*. Boston, MA: Federal Reserve Bank.

Ruegg, R. 2001. "Taking a Step Back: An Early Results Overview of Fifty ATP Awards." In National Research Council. The Advanced Technology Program: Assessing Outcomes. Charles W. Wessner, ed. Washington, D.C.: National Academy Press.

Ruegg, R. 2003. Interview of R. Wesson. December 1. National Science Foundation, Arlington, VA.

Ruegg, R. 2003. Interview of R. Coryell. October 23. National Science Foundation, Arlington, VA.

Ruegg, R. 2004. Interview of C. Albus. January 7. National Science Foundation, Arlington, VA.

Ruegg, R. 2005. Interview of J. Hennessey. October 18. National Science Foundation, Arlington, VA.

Ruegg, R. 2006. Interview of J. Hennessey. March 3. National Science Foundation, Arlington, VA.

Ruegg, Rosalie, and Irwin Feller. 2003. *A Toolkit for Evaluating Public R&D Investment Models, Methods, and Findings from ATP's First Decade*. NIST GCR 03-857.

Ruegg, Rosalie, and Patrick Thomas. 2007. *Linkages from DoE's Vehicle Technologies R&D in Advanced Energy Storage to Hybrid Electric Vehicles, Plug-in Hybrid Electric Vehicles, and Electric Vehicles*. U.S. Department of Energy/Office of Energy Efficiency and Renewable Energy.

Sahlman, W. A. 1990. "The structure and governance of venture capital organizations." *Journal of Financial Economics* 27:473–521.

Saxenian, Annalee. 1994. *Regional Advantage: Culture and Competition in Silicon Valley and Route 128*. Cambridge, MA: Harvard University Press.

SBIR World. SBIR World: A World of Opportunities. Available online at <http://www.sbirworld.com>.

Scherer, F. M. 1970. *Industrial Market Structure and Economic Performance*. New York: Rand McNally College Publishing.

Schumpeter, J. 1950. *Capitalism, Socialism, and Democracy*. New York: Harper and Row.

Scott, John T. 1998. "Financing and leveraging public/private partnerships: The hurdle-lowering auction." *STI Review* 23:67–84.

Small Business Administration. 1992. *Results of Three-Year Commercialization Study of the SBIR Program*. Washington, DC: U.S. Government Printing Office.

Small Business Administration. 1994. *Small Business Innovation Development Act: Tenth-Year Results*. Washington, DC: U.S. Government Printing Office (and earlier years).

Sohl, Jeffrey. 1999. *Venture Capital* 1(2).

Sohl, Jeffery, John Freear, and W.E. Wetzel Jr. 2002. "Angles on Angels: Financing Technology-Based Ventures—An Historical Perspective." *Venture Capital: An International Journal of Entrepreneurial Finance* 4 (4).

Stiglitz, J. E., and A. Weiss. 1981. "Credit rationing in markets with incomplete information." *American Economic Review* 71:393–409.

Stowsky, J. 1996. "Politics and Policy: The Technology Reinvestment Program and the Dilemmas of Dual Use." Mimeo. University of California.

Tassey, Gregory. 1997. *The Economics of R&D Policy*. Westport, CT: Quorum Books.

Tirman, John. 1984. *The Militarization of High Technology*. Cambridge, MA: Ballinger.

Tyson, Laura, Tea Petrin, and Halsey Rogers. 1994. "Promoting entrepreneurship in Eastern Europe." *Small Business Economics* 6:165–184.

U.S. Congress, House Committee on Science, Space, and Technology. 1992. *SBIR and Commercialization: Hearing Before the Subcommittee on Technology and Competitiveness of the House Committee on Science, Space, and Technology, on the Small Business Innovation Research [SBIR] Program*. Testimony of James A. Block, President of Creare, Inc. Pp. 356–361.

U.S. Congress. House Committee on Small Business. Subcommittee on Workforce, Empowerment, and Government Programs. 2005. *The Small Business Innovation Research Program: Opening Doors to New Technology*. Testimony by Joseph Hennessey. 109th Cong., 1st sess., November 8.

U.S. Congress. Senate Committee on Small Business. 1981. Small Business Research Act of 1981. S.R. 194, 97th Congress.

U.S. Congressional Budget Office. 1985. *Federal financial support for high-technology industries*. Washington, DC: U.S. Congressional Budget Office.

U.S. General Accounting Office. 1987. *Federal research: Small Business Innovation Research participants give program high marks*. Washington, DC: U.S. General Accounting Office.

U.S. General Accounting Office. 1989. *Federal Research: Assessment of Small Business Innovation Research Program*. Washington, DC: U.S. General Accounting Office.

U.S. General Accounting Office. 1992. *Small Business Innovation Research Program Shows Success but Can Be Strengthened*. RCED–92–32. Washington, DC: U.S. General Accounting Office.

U.S. General Accounting Office. 1997. *Federal Research: DoD's Small Business Innovation Research Program*. RCED–97–122, Washington, DC: U.S. General Accounting Office.

U. S. General Accounting Office. 1998. *Federal Research: Observations on the Small Business Innovation Research Program*. RCED–98–132. Washington, DC: U.S. General Accounting Office.

U.S. General Accounting Office. 1999. *Federal Research: Evaluations of Small Business Innovation Research Can Be Strengthened*. RCED–99–198, Washington, DC: U.S. General Accounting Office.

U.S. Public Law 106-554, Appendix I–H.R. 5667, Section 108.

U.S. Senate Committee on Small Business. 1981. Senate Report 97–194. *Small Business Research Act of 1981*. September 25. Washington, DC: U.S. Government Printing Office.

U.S. Senate Committee on Small Business. 1999. Senate Report 106–330. *Small Business Innovation Research (SBIR) Program*. August 4. Washington, DC: U.S. Government Printing Office.

U.S. Small Business Administration. 1994. *Small Business Innovation Development Act: Tenth-Year Results*. Washington, DC: U.S. Government Printing Office.

U.S. Small Business Administration. 2003. "Small Business by the Numbers." SBA Office of Advocacy. May.

Venture Economics. 1988. *Exiting Venture Capital Investments*. Wellesley, MA: Venture Economics.

Venture Economics. 1996. "Special Report: Rose-colored asset class." *Venture Capital Journal* 36 (July):32–34 (and earlier years).

VentureOne. 1997. National Venture Capital Association 1996 annual report. San Francisco: VentureOne.

Wallsten, S. J. 1996. The Small Business Innovation Research Program: Encouraging Technological Innovation and Commercialization in Small Firms. Unpublished working paper. Stanford University.

Wessner, Charles W. 2004. *Partnering Against Terrorism*. Washington, DC: The National Academies Press.